Heroes Without Legacy

Heroes Without Legacy

American Airwomen, 1912–1944

by
Dean Jaros

UNIVERSITY PRESS OF COLORADO

Copyright © 1993 by the University Press of Colorado
Published by the University Press of Colorado, P.O. Box 849, Niwot, Colorado 80544

The University Press of Colorado is a cooperative publishing enterprise supported, in part, by Adams State College, Colorado State University, Fort Lewis College, Mesa State College, Metropolitan State College of Denver, University of Colorado, University of Northern Colorado, University of Southern Colorado, and Western State College of Colorado.

Cover: Gladys O'Donnell, winner of many air races, c. 1930.
Smithsonian Institution Photo No. 73-5629.

Library of Congress Cataloging-in-Publication Data
Jaros, Dean
  Heroes without Legacy : American airwomen, 1912–1944 / by Dean Jaros.
  p. cm.
  Includes bibliographical references and index.
  ISBN 0-87081-312-9 (alk. paper)
  1. Women air pilots—United States—Biography. I. Title.
TL139.J37 1993
629.13'092'273—dc20
[B]   93–42989
                                                                              CIP

The paper used in this publication meets the minimum requirements of the American National Standard for Information Sciences—Permanence of Paper for Printed Library Materials.
ANSI Z39.48–1984

∞

10 9 8 7 6 5 4 3 2 1

Contents

Illustrations	vii
Acknowledgments	ix
Introduction	3
1. Women Conquer the Air: Promise Demonstrated	9
2. Comments on the Conquest: Promise Proclaimed	69
3. The Decade After World War II: Promise Denied	115
4. The Long Term: Promise Forgotten	149
5. Explanations: Promise Unrealized	183
6. Belated Legacy: Promise Reborn	217
Glossary	239
References	243
Index	255

Illustrations

Harriet Quimby and the Moisant International Aviators   17

Ruth Law and Her Exhibition Airplane   19

Mabel Cody in an Airshow Act   25

Phoebe Omlie in an Aerial Political Campaign   35

Louise Thaden After Establishing an Endurance Record   41

Amelia Earhart and Her Instructor   45

Jacqueline Cochran After Winning the Bendix Trophy Race   51

Women Airforce Service Pilots   62

Amelia Earhart   76

Harriet Quimby   83

Katherine Stinson   85

Ruth Elder   88

Fay Gillis Wells in the Comics   107

Women Airforce Service Pilot   126

Nancy Harkness Love   128

Helen Richey   130

Jacqueline Cochran   139

Acknowledgments

A number of people aided in the preparation of this book. Those who supplied information, facilitated access, or provided other direct help include: Ruth Alexander, Colorado State University; Peggy Baty, Parks College; Betty Shea Boyd, Alameda, California; Amy Carmien, Women in Aviation, Inc.; Deborah Douglas, National Air and Space Museum; Rigdon Edwards, Sweetwater, Texas; Barbara Lynch, Rockville, Maryland; Dorothy R. Niekamp, Indiana University; Yvonne C. Pateman, Women Airforce Service Pilots; Charlene Sammis-Singleton, International Society of Women Airline Pilots; Beatrice Steadman, International Women's Air and Space Museum; and Cheryl J. Young, St. Paul, Minnesota. Many of these individuals are cited or otherwise discussed in the text.

I owe a great deal to the personnel of the following libraries where I did much of my research: Colorado State University, Ft. Collins, Colorado; Experimental Aircraft Association, Oshkosh, Wisconsin; National Air and Space Museum, Washington, D.C.; The Ninety-Nines, Inc., Oklahoma City, Oklahoma; Northrop University, Los Angeles, California.

The staff of the Graduate School at Colorado State University accepted with good humor the additional work that the writing of this book imposed on them. Not only did they maintain their usual high level of performance on their regular tasks, but they tolerated with much understanding a sometimes rather strung-out demeanor on the part of their dean. Special thanks are due to Janie Richmond who managed a sizable information flow throughout and whose preparation of the manuscript is greatly appreciated.

Provost Elnora M. Gilfoyle was extremely understanding of my desire to write this book. She permitted a flexible approach to my regular responsibilities so that time, energy, and tangible resources were available for this effort. I am deeply grateful.

Finally, my two daughters, Cindy and Lisa, in the course of achieving professional success, and quite unconsciously on their part, sensitized me to some of the broad issues that underlie this book. I am delighted to have learned from the younger generation.

Heroes Without Legacy

Introduction

To paraphrase Rousseau, "Woman is born equal to man; yet everywhere she is in chains." Throughout most of recorded history, few people would have agreed with this statement. Indeed, whether it is true or false would rarely even have been a topic for discussion; the whole issue would have been regarded as senseless. Everyone — men and women alike — "knew" that women were inferior and so it was only natural that they should be restricted to second-class citizenship. This view, dominant for centuries, has foreclosed serious consideration of what women could do or might desire to do. Relatively few women have been able to seize opportunity, and advocates of gender equality have remained, at best, on the fringes of the culture. Philosophers and scholars, if they allowed such notions to cross their minds, generally have confirmed feminine weakness and justified women's low social status. Even thoughtful and sincere observers find it difficult to deal with the tonnage of tradition.

Nonetheless, there have been some recent stirrings. The modern age appears to be a bit more "empirical" than previous eras. People may be more "scientific" than in past millennia; they may pay more attention to facts. What they can see may be more important than old customs and articles of faith. Accordingly, in the last century or so — and particularly in recent decades — evidence has been accumulating. This evidence has shown women's abilities and potentials, including those in traditionally masculine arenas, to be fully equal to those of men. In a word, women are not inferior at all.

To be sure, not everyone has accepted this. New truths are disturbing to comfortable social inertia. When those truths call into question something very basic to the culture, like gender relationships, they are likely to cause real anxiety. Hence, many people have ignored evidence of women's abilities, and more than a few have hotly denied it. Historically validated practice has great staying power even in the absence of a strong factual base.

But despite ancient, established beliefs, opposition to sexism is growing. The motivation, of course, is not entirely out of respect for empirical evidence. Many individuals find gender stereotypes simply unjust, and others deplore the great social inefficiency that they impose on society. At an even more basic level, women in the twentieth century have become much better informed as communications technology has mushroomed. At the same time, women have increased their overall level of education at a faster rate than their male counterparts. Inevitably, this has led to a whole new set of well-expressed political demands because women are no longer inclined to accept second-class citizenship.

So we have come to a time of much conflict, and just possibly historically significant conflict at that. The decade ahead is likely to be crucial. To be sure, more evidence and more activism will further discredit negative images of women. But those very developments increase the overall threat to society's mores, and a reaction or backlash is thus in store as well.

The purpose of this book is to join the fight for full citizenship for women. By analyzing events from another era, events that might have led to significant and permanent advances, it attempts to demonstrate why, in a particular historical situation, gains were not forthcoming despite a seemingly very favorable set of conditions. In this are lessons for today.

This book is about a failure in persuasion. Social change is sometimes driven by military conquest, the collapse of economies, or even huge natural disasters. But at least in Western democracies, it is also a matter of opinion — public opinion, the opinion of government decision makers, the opinion of business leaders, and of course the opinion of those who stand to gain or lose. For change to occur, the presently hostile must be convinced to reexamine the issues; the potentially supportive must be mobilized; the enthusiastic must be focused and directed. Unfortunately, at several propitious moments for gender relationships in this country, these are exactly the things that did not happen.

In the 1890s, a large number of women were entering the medical profession. Enrollments in medical schools were substantially female and the proportion of female students was rising. It would not have been unreasonable to project twentieth-century medicine as something of a women's career. But with dramatic suddenness, the trend was aborted; medical school enrollments reverted to being almost completely male and women's practice remained very restricted. What led so many women to explore medicine at this time? Who or what ultimately undermined them? Were there identifiable, active enemies or were these women defeated by broader social and economic trends? Why were female medical professionals and aspirants unable to resist these forces? What long-term impact did these early female physicians have? Did their pioneering efforts inspire anyone? Could this scenario have ended differently? Answers to such questions provide insights for intervention in parallel events that are happening today.

Similarly, there are lessons for us in the study of American women in the early days of aviation. This book is about a series of female pilots who, between 1912 and 1944, compiled a remarkable list of aviation achievements. I call them "airwomen," a convenient shorthand reference that is also a descriptively accurate term. Their accomplishments during this thirty-two–year period are filled with drama and excitement. Women competing head-to-head with men won transcontinental air races. Female pilots often exceeded their male counterparts in derring-do, performing spectacular if foolish feats such as flying under Manhattan bridges — to the delight of a public less jaded than today's. Both women and men completed endurance flights that lasted for weeks on end, and pilots of both genders visited the four corners of the earth.

In addition, during this period aviation began to mature in a technical and commercial sense — and women were among the entrepreneurs and innovators. They participated in the establishment of transportation routes and the transition from open cockpit biplanes to machines with practical cabins. Both men and women popularized flying, as aviation correspondents to major magazines and as authors of best-selling books. The press was greatly impressed by female pilots and proclaimed their ability and worth.

Public fascination with flying was so intense that it developed into a nationwide love affair with the airplane — a press-mediated euphoria that culminated in the nearly universal expectation of the dawn of a new era based on flight. The air age would be prosperous, aesthetically sophisticated, humane, and above all characterized by universal participation and cooperation among peoples.

These decades also marked a time of more general social experimentation — this period spans Progressivism, Prohibition, and the New Deal. The airwomen were contemporaries of the "flappers," though many of them would not have wanted to be identified as the latter. These were also years of economic instability, international tensions, and war, forces that directed women toward roles never before available to them. Some airwomen fully understood the potential implications of their accomplishments for women's role in society, and they wrote articles calling for a more gender-equitable culture. In short, the excitement of innovation and change was in the air. Most people were optimistic about the coming decades. It did not take much imagination to foresee a technologically advanced future in which women were key players.

While the future seemed bright with promise, ultimate reality turned out to be harsh and bleak. There were pioneering female transport pilots in the 1930s — one of them even flew for the scheduled airlines — but half a century was to pass before airliner cockpits were even slightly open to women. In the early days of World War II the Women Airforce Service Pilots flew the newest fighters and bombers on a routine, albeit noncombat, basis, but the disbanding of the WASPs in 1944 signalled the exclusion of women from military flying for a period almost as long.

Leaders among the female pilots retreated: some pursued individual recognition even at the cost of women aviators generally; others took postures designed specifically to avoid posing a threat to men; still others left aviation altogether. Until very recently, female pilots of this period virtually disappeared from public view, and all but one — Amelia Earhart, a creation of the media industry — vanished from literature and popular culture. The airwomen are nearly absent from aviation history. Serious advocates of gender-role transformation ignored these women pilots altogether. Scholars of women's history similarly paid little attention to their accomplishments. Even their professional descendants, commercial female pilots, do not look to the airwomen for inspiration.

To be sure, the factors affecting women's aviation were not isolated from more general social forces. By 1944 it was clear that the golden promises of aviation had not been fulfilled. Indeed, the anticipated aerial age had failed in almost every respect. A helicopter or roadable aircraft in every American garage remained a fond wish. Nations were not peacefully united by mutual understanding stemming from easy and frequent worldwide air travel; instead, airplanes were powerful implements in a terrible, ongoing war. Broad trends in the way American women viewed themselves — and in concepts of feminine self-realization — buffeted society, spin-offs of dramatic economic and political events that shaped all of history. Nonetheless, as this book will demonstrate, it is appropriate to consider the airwomen an important part of the social dynamics of this era rather than a derivative historical curiosity.

Like their sisters who sought medical degrees, the airwomen stepped far beyond contemporary expectations for women's behavior. It is tempting simply to describe the lives of these intrepid but neglected individuals. After all, knowledge of these lives may translate into concrete support for the heroes' underlying principles. The adventures of the airwomen have, however, been recounted by a number of very competent writers. Although neither the histories nor the biographies of airwomen are as comprehensive as those for men of similar accomplishment, notable works continue to appear and important knowledge gaps are slowly being closed. I take the heroism of these airwomen to be a matter of record and will discuss their substantive accomplishments only enough to confirm their significance.

In a similar way, it would be easy to become an evangelist for the airwomen. But this purpose too must be declined. Even though this book unapologetically admires the airwomen and even though right in the title it proclaims them heroes, it is most emphatically *not* an exercise in hero-worship.

Along with the developing body of history and biography, there is a growing literature of adulation. Particular authors, and even entire organizations, have become devoted to the memories of individual airwomen. Others consider the airwomen above criticism. To be sure, adulation is clearly deserved and can perhaps lead to support for the values for which the airwomen stood. However, uncritical praise can easily lead to satisfaction and complacency: to affirm greatness then becomes an end in itself and no further action is deemed necessary.

No doubt some authors believe that the shining images they proffer will secure their airwomen-heroes an appropriate place in history. Nothing could be further from the truth. The long-term contribution of the airwomen will emerge only from demanding analysis. The credibility that comes from critical assessment is vital to any argument. If their stories and principles are to be important for the future, we must not only celebrate the airwomen's stunning successes, but also acknowledge their shortcomings.

This book, therefore, focuses less on great deeds as such and more on context and consequences. It examines basic features of the period between 1912 and 1944 and tries to assess reactions to the airwomen. The short-term result of these exciting times as well as their long-term implications — or lack thereof — are also presented. *Heroes Without Legacy* offers some thoughts on why the airwomen did not have a much greater impact and, finally, considers whether an airwomen's legacy might yet develop despite the intervening years. Rather than popularize or glorify the airwomen, it provides an interpretive assessment of them, their times, and their broader meaning. This assessment conveys two messages. One — of courage, accomplishment, and pride — is vindicated if not widely appreciated. The other — of dashed hope and failed promise — carries the sting of disappointment and perhaps even implies reproach — but it also carries lessons of value and these are the goal.

✼ 1 ✼

Women Conquer the Air:
Promise Demonstrated

Since prehistoric times, human flight had been the stuff of dreams, magic, fantasy, the supernatural, the religious, the unattainable. Most cultures have had some sort of myth in which people, or gods who take the form of people, have the ability to fly. But until this century none has had an airborne reality. The invention of aviation was thus truly revolutionary. Dimensions of power and privilege previously reserved for the imagination were suddenly hard facts. Allegory and experience seemed to change places. Flying was mind-boggling for the American public, who greeted it with astonishment, wonder, or disbelief. Whether due to aeons of universal tradition or something inherent in the human psyche, the advent of airplanes caused something of a cognitive crisis. For most people, it was very difficult to come to grips with the idea of actually leaving the ground, to say nothing of a sustained voyage above the earth.

Two Dayton bicycle mechanics first flew in 1903, but recognition of their feat came grudgingly. For at least five years, the great majority of Americans were unaware of the existence of the airplane. Most reporters initially thought Orville and Wilbur Wright were cranks and did not bother to investigate their claims. Such early newspaper stories as did appear were simply rejected by the public — flight was too fantastic an idea to be credible. The first published, eyewitness account of an airplane flight did not come in the national press; indeed, it was not even in a local newspaper. Incredibly, what should have been a great reporting scoop appeared in an obscure, specialized journal, *Gleanings in Bee Culture* (Root 1905). In 1904 an Ohio honey producer dropped by the Wrights' establishment and watched the brothers do a few circuits around the field. No one was interested in his attempt to tell the world about it, and so he resorted to the little apiarian magazine that he published himself. Unless we allow that airplanes and bees should be classed together as flying objects, this was an incongruous — not to mention inauspicious — journalistic beginning.

Even though press attention gradually developed, people often would not believe in airplanes until they witnessed them with their own eyes. Observers at early airshows, now unable to escape reality, were left literally speechless (see examples in Bilstein 1978). Flight apparently resonated at a very deep emotional level.

Once this mental dam was breached, however, the flood of "airmindedness" was of huge proportions. The public emotions that at first acted as a constraint now fueled enthusiasm and celebration. The wave of aviation euphoria that swept the country took on faintly religious overtones; Joseph Corn called it "the winged gospel" (1983). Perhaps it is not too much to speak of a widespread conversion experience in the United States.

The fascination with flight led inevitably to much aerial showmanship. Aerial exhibitions became very popular, perhaps because extensive public demonstrations were necessary to convince the public of the existence of airplanes. Among the masses, hesitancy gave way to credulity and full-fledged airshows were a reality by 1910. A huge spectator demand developed and plenty of adventuresome spirits — moved by idealistic airmindedness or more likely the lure of money — were available to meet the demand.

As a result, much of early aviation was geared toward entertainment. The public thrilled to the loud noise and bright colors of early machines, to races and aerobatics, to aerial trapeze acts, wing walking, and any maneuver that was unusual or appeared to be dangerous. Even an ordinary parachute jump or a few loops by a single plane could be a source of great excitement in rural areas. Though aviation and the public have changed a great deal in ninety years, the same dynamics are still visible in many local airshows and especially in the crowds drawn to the Experimental Aircraft Association's annual fly-in at Oshkosh, Wisconsin.

Though showmanship was certainly the most visible early manifestation of airmindedness, it was not the only one. To most Americans of that era — the public as well as aviators themselves — aviation was very grand indeed; they believed that something of long-term social significance was at stake. The real meaning of aviation was in the betterment of humankind. Though Americans loved personalities and performances, the miracles of aviation portended more than a chapter in circus history.

This expectation proved true, though not quite in the fashion that many anticipated. Throughout most of the first half-century of flight, journalists and armchair philosophers freely projected utopian social outcomes based on aviation technology. The wildest asserted health benefits from being aloft and even the eventual development of a superhuman species. More restrained futurists predicted a decline in regional divisiveness, the end of oppressive railway monopolies, and mass escape from life in "corrupt" cities (Bilstein 1983, 158–163; Corn 1983, 29–50). Though such unbridled optimism seems in retrospect naive, aviation technology did advance in spectacular fashion. However, aviation's greatest impact resulted from the application of that technology to more traditional concerns.

In the earliest years, few business titans, senior military officers, or established scientists hailed the flying machine. Such practical people regarded it as little more than an amusing curiosity. This quickly changed, however, as people with vision began to see the possibilities. Not only was aviation miraculous in some deep-seated psychological sense, but it was likely to have near-miraculous, tangible consequences. Conscious promotion began; aviation grew to become a sophisticated travel medium and a mechanism important in a host of commercial, military, and scientific pursuits. Entrepreneurs saw the beginnings of a widespread transportation industry. Generals (with delight) and admirals (with dismay) soon discovered that aerial bombs could indeed sink ships. Explorers and prospectors found that their scope of operations could be extended from hundreds to tens of thousands of square miles.

In short, the earliest history of aviation caused it to have a dual personality. On the one hand, it was an expression of showmanship, an entertainment medium driven by adventurous pilots and thrill-seeking customers. This was the performance dimension. On the other hand, it was an expression of a search for meaning, for applications of a new technology that would improve daily life and address the long-standing business, military, and scientific concerns of the nation. This was the practical dimension.

Of course these two dimensions were not always antithetical or even, in a given event, separate from one another. Indeed, in an era prior to systematic testing, the experience derived from the performance of ever new and more bizarre stunts helped to provide a knowledge base for improvements in both airplanes and engines. Purveyors of aeronautical entertainment argued that the heightened consciousness of aviation they promoted among the public would benefit the economy and enhance military preparedness. Many aviation heroes were both performers and technocrats. Charles Lindbergh, perhaps the most famous of them all, began as a midwestern barnstormer but spent most of his career promoting commercial and military aviation. Nonetheless, after about 1920 both observers and participants understood this duality, if only dimly.

Against this backdrop, women were able to conquer the air. As this chapter demonstrates, the airwomen compiled an impressive record both of performance and of practical contribution. In some ways, the period between 1912 and 1944 does not seem terribly favorable to women. In almost all areas, their opportunities were much more limited than today's. Male-dominated culture remained, for the most part, well established. Even the successful reform movements of the period were a mixed blessing. Near the turn of the century, the historical thrust known as Progressivism gained a great deal of political force. It focused on the many abuses of early industrial capitalism and the social consequences of unrestrained corporate greed. Prominent in its platform was the improvement of the condition of women. However, legislation of the Progressive era focused on women's special roles as childbearers and parents, on their particular "weakness" and need for special protection. Laws limiting the number of hours women could work in factories were typical of this era. While such an orientation may have made sense given the social realities of the time, it did not promote equity between men and women nor did it encourage ambition or achievement among women. The prototypical Progressive woman was healthy and free from exploitation, to be sure, but she was at home caring for her children.

These conditions hardly seem to provide a template for the development of a hard-driving, success-oriented group such as the airwomen. How then did they emerge? The answer seems to be twofold. First, although this was not a time of revolutionary attack on male dominance, change was in the wings. Progressivism, limited though it was, stimulated a good deal of diffuse thought; many were uneasy with their current social reality even if they had little concrete vision of the future that might replace it. Though there was no ready-made agenda for the airwomen, times were ripe for them to create one of their own.

Second, the peculiar features of early aviation itself gave the airwomen advantages that their sisters in other endeavors, even if equally ambitious, were unable to match. Aviation's strong performance dimension is perhaps the most obvious. Even in traditional male-dominated culture, there has usually been a place for female performers. Theater and dance come readily to mind and, to a lesser degree, music. Perhaps being a record-setting female pilot was no more shocking than being an actress or a singer — or any other kind of entertainer. Resistance to women in aviation, at least initially, may thus have been inherently less than that in established professions like the law.

It is surely significant that there were many nonpilot female performers in early airshows. Their work was parallel to the Ziegfeld Follies, common burlesque, or the circus, all popular — and extremely sexist — forms of entertainment at the time. Like a Ziegfeld Girl, the typical aviation performer was physically attractive; like her circus counterpart, she was willing to appear in various acts that either involved danger or simulated it.

Diminutive stature, general adolescent appearance, very feminine clothing, and the ability to look helpless led to the promotion of Tiny Broadwick. Though she began with low-budget travelling shows, she may have been the first woman in history to make a parachute jump — from a tethered balloon in about 1910. By 1913 she had joined early aviation pioneer Glenn Martin as the major parachutist and handbill feature in his airshows. Broadwick's career was preceded by a colorful if not bizarre youth and in addition she enjoyed a very long life. She thus made good human-interest interview material and enjoyed brief renewed attention in the seventies (Vecsey and Dade 1979, 30–39).

Margie Hobbs, alias Ethel Dare, the Flying Witch, did many in-flight interairplane transfers by means of rope ladders in 1919 and 1920. She was a former circus trapeze artist who well understood the draw of scanty attire and well-posed photographs. Among the best-known stuntwomen was Lillian Boyer who, beginning in the early twenties, variously attached herself to or leapt from flying planes. Dangling from special fittings, landing gear, or wing struts, she claimed to be "hanging around" the airport. Gladys Roy danced the Charleston on the top wing of a biplane in the late twenties. She was not much different from many similar performers, except that she was killed by walking into the spinning propeller of a plane idling on the flight line. When Helen Lach, a 1927 airshow parachutist, appeared in shorts, a halter top, and a flying helmet, not only her profession but also her costume appeared quite daring by contemporary standards (Brooks-Pazmany 1983, 4–7). This list could be extended greatly in terms of both numbers of people and scope of time. Indeed, this performance genre survives in contemporary female airshow artists who, in far greater numbers than their male counterparts, continue to cavort on the various surfaces and in the rigging of aerobatic biplanes.

Clearly these performers did not have traditional lifestyles nor were they lacking in physical courage; perhaps on those accounts they deserve some respect and attention. But in the larger scope of things they were unremarkable, for use of women in similar performance roles has been fully within the tradition of male-dominated culture. In the minds of many, this is not accomplishment at all, but instead continued sexual exploitation for the indulgence of the masses.

What is notable about the airwomen, even if one considers only their role as performers, is that they projected themselves far beyond the Broadwicks and the Boyers. Tradition may have provided an entree to passive roles where cute and vulnerable "girls" were acted upon by the raw, untamed forces of flight, but many women boldly went right on by to the cockpit, literally seizing the controls. Perhaps they only extended a tradition, but they did so until it cracked.

Turning to the practical dimension of early aviation, we see another factor that may have facilitated the participation of women. Traditionally, business, military affairs, and science have been masculine domains. Developing technologies likewise have been exclusive male preserves. But as we saw above, aviation was not merely another technological development; it was a revolutionary phenomenon from the moment of its inception. Since it was miraculous and vaguely supernatural, who knew what rules, regulations, and practices should govern it? In most revolutions, new elites appear. Lineage and custom become somewhat less important under these conditions. At least until older, established elites recover from their surprise, it may be possible for newcomers to get in on the ground floor.

If opportunity knocked in this way, the airwomen certainly responded. Unsatisfied with the environments in which they found themselves, they took the new technology into their own hands. The combination of piloting skill and imagination led them not only to push performance to the maximum but also to a more practical kind of achievement, one associated with the intellect, abstraction, research, and substantive contributions to humankind.

Performance and Practicality Among the First Airwomen

On April 16, 1912, at Dover, Harriet Quimby, the first American woman to hold a pilot's license, calmly mounted to the cockpit of a fifty-horsepower monoplane and flew across the English Channel. This was the first crossing by a female pilot and the event generated a great deal of excitement on that account alone. A female pilot, though not unique, was certainly unusual; one like Quimby, who had excellent interpersonal skills and a flair for the dramatic, was lionized by the press. A writer for *Leslie's Weekly* and an accomplished drama critic, she well understood both performance and public relations. Her specially designed, hooded purple satin flying suit, for example, predated some of the attention-getting garb that pilots of both sexes were to affect in the next decade.

The citizens of Hardelot, Quimby's landing site on the French coast, carried her on their shoulders and cheering crowds welcomed her in Paris. She might very well have become a much greater public figure but for some bad timing. The *Titanic* tangled with an iceberg on the very day of her flight, and it crowded her off the front page of newspapers worldwide. Her return to the United States similarly coincided with a massive suffrage demonstration in New York, so none of the reporters who customarily interviewed women were at the dock to greet her.

Nonetheless, charming anecdotes — focusing on her gender — survive. Gustav Hamel, underestimating Quimby's skill, offered to participate in a Channel flight ruse. He would don her purple flying suit, fly across the Channel to an isolated site, return airplane and attire to the waiting Quimby, and disappear before press and public arrived. She declined his chivalrous but dishonest suggestion (George 1980; Gwinn-Jones 1984; Powell 1982).

Is the only significance of Quimby's flight the fact that it was accomplished by a "special category" of pilot? After all, Louis Bleriot had flown across the Channel in 1909; Quimby was not first. But flights across the Channel were not common in 1912, regardless of who the pilots might be. Bleriot's feat occurred only three years previously and there had been few intervening aerial voyages between England and France. Quimby's flight, the first in anything other than clear weather, would have been notable regardless of the gender of the person at the controls. The feasibility of air travel was further demonstrated; the mechanical reliability of airplanes was in some small degree confirmed; the need for advances in navigational equipment was made more clear. Hers was indeed a contribution, in an absolute sense, to the technical development of aviation.

That Quimby was a leader in the development of aviation quite apart from any considerations of gender becomes clear if one looks beyond the drama of the Channel. In 1911, as part of the mixed-gender Moisant exhibition team, Quimby helped introduce aviation to Mexico, participating in some of the earliest flights there as part of a presidential inauguration ceremony. After the Channel flight, Quimby remained active in various pioneering aviation events. Even in death she left an important message for all of aviation. In 1912, Quimby and a passenger died as her new white airplane, acquired on her recent trip to France, plunged into Boston harbor. The crash was one of the first to be carefully analyzed. The fact that both victims were thrown from the plane occasioned considerable discussion of

Harriet Quimby (third from the right) and the Moisant International Aviators. Mathilde Moisant, another female member of the team, is also pictured, c. 1912. Smithsonian Institution Photo No. A 44401 C.

the use of seatbelts; prior to that time the utility of this safety feature was only dimly understood.

Although aviation surely provided the public with tremendous vicarious adventure, not all of its contributors were shimmering personalities and headline material. Not surprisingly, this is true of women as well as men. A good example is Bernetta Miller. A contemporary of Quimby's who also worked with Moisant International Aviators, Miller had no notable public performance record at all. The Moisant group was somewhat unusual in that its preferred machine was the monoplane. Indeed, the far more prevalent biplane was to remain supreme for many years, but advanced thinkers, including some in the U.S. Army, had begun to wonder about potential benefits of a single wing. Though they would not immediately be vindicated, these Army experts called on the Moisants to provide a demonstration of the monoplane at College Park, Maryland, in October of 1912. Not only was Miller part of the demonstration team, but she also had the honor of doing the very first monoplane demonstration flight for U.S. officials (Oakes 1978, 20–21). This was a small though important step in the adoption of a new technology.

Ruth Law was perhaps the most ambitious of the early airwomen. Her performance record, with regard to both entertainment and setting records, was spectacular. But Law quickly grasped the commercial potential of aerial demonstration and became the first aviation businesswoman. She established a successful passenger-carrying and exhibition operation as early as 1913. She worked resort areas in the Northeast and Florida; reportedly, she did quite well financially. Law was also known for nighttime performances featuring multiple loops illuminated by ground-operated spotlights. As part of the 1916 ceremony initiating electric lighting of the Statue of Liberty, she flew several circuits around the monument with the word "Liberty" in lights on the bottom of the wing.

Not only did Law expand aircraft use in business at night, she was also an airmail pioneer, making the very first delivery to Manila in 1919. Probably her most important achievement was a spectacular Chicago–New York flight in 1916. Cross-country flying had never before been so convincingly demonstrated. In the process, Law set a nonstop distance record, from Chicago to a point near Elmira, of 590 miles, eclipsing Victor Carlstrom's previous high of 452 miles. Law's accomplishment was very well received by government officials, the economic elite, and the press. Dinners in her honor were attended by Admiral Perry, several cabinet members, and President Wilson himself (Adams and Kimball 1942; "Ruth Law's Record Breaking Flight," 1916; "Miss Ruth Law Presented With $2,500 Purse," 1917; Gee 1932).

The record-breaking trip also demonstrated some of Law's technical contributions. In order to prepare the stock Curtiss Pusher airplane for such a demanding trip, she designed a number of modifications including larger fuel tanks and protection from the elements. The most interesting of Law's innovations was a chart case. A rolled map, protected by a glass cover, could be scrolled in such a way as to progressively reveal the flight path along the route. The advantages over trying to read and fold large paper maps — especially in early, windy airplanes — are obvious. Variations of this device were to become common.

Law, who unsuccessfully sought a role as a combat pilot in World War I, showed the world new potentials for aviation. Although many men

Ruth Law in her Curtiss Model E biplane flies an exhibition routine, c. 1915. Smithsonian Institution Photo No. 92-7959.

and women were doing the same during these years, Law must be regarded as among the best.

The family name "Stinson" has a long and distinguished history in American aviation. The Stinson firm was one of the major U.S. aircraft manufacturers for a period of nearly forty years. In the twenties, Stinson was among the companies pioneering large, cabin monoplanes. This was an extremely important development in that the open-cockpit biplanes, which tenaciously persisted as the standard configuration until long after World War I, were unsuited to many sophisticated commercial functions. This was particularly true with regard to passenger service, where cramped seating and exposure to the elements deterred many potential riders. Stinson production evolved multi-engine airliners including a sleek, all-metal low-wing tri-motor in the mid-thirties. A prominent line of general aviation aircraft continued into the fifties, but the firm merged with a larger corporation when the post–World War II aviation boom failed to develop.

The best-known figure associated with the company was Edward "Eddie" Stinson, a barnstormer who became a designer, entrepreneur, and ultimately corporate tycoon. It was not Eddie, however, who founded the Stinson family dynasty, nor was it he who had the most distinguished flying record. Those honors belong to Katherine and Marjorie Stinson, who taught their brother to fly.

Katherine Stinson was a marvelously ambitious woman, who quickly recognized the financial benefits of aviation. As a youngster with aspirations for the concert stage, she was impressed by the earning potential of exhibition flying: she believed a temporary stint as a professional pilot would finance the purchase of a piano, training, and other underpinnings of a musical career. She earned a pilot's license in 1912 and in 1913 formally established a flying business with her mother. Katherine did a great deal of exhibition flying in 1913, eventually settling in San Antonio — where a general aviation airport is today called Stinson Field.

Katherine soon abandoned her plans to be a pianist and concentrated on the business of flying. Marjorie joined the firm as second pilot and the mother-daughter trio founded a flightschool in 1915. The all-female ownership and management of such a facility was unusual in those days, as was its level of success. World War I created a demand for pilot training, and the Stinson school's reputation was such that the military sought out its instructors. Many American pilots trained there, but the majority of military students were from the Royal Canadian Flying Corps. Upon graduation, they were posted to the lines in France. Marjorie in particular was the object of much appreciation, earning the appellation "flying schoolmarm." Her Canadian combat pilot trainees at the front were informally known as the "Texas Escadrille" — after the Lafayette Escadrille, the American volunteer pilots group that served with the French forces in the early years of World War I. Ironically, the school had to close when the government imposed severe restrictions on civilian flying following the entry of the United States into the war.

The flightschool business did not keep Katherine from an extensive exhibition schedule. In 1915 she became the first woman to perform a loop and shortly after introduced her own aerobatic innovation, which she called the "dippy twist." It involves performing a roll at the top of the loop, such that the plane, inverted after ascending to the very apex, quickly attains an upright attitude and then returns to inverted before the descent down the back side. This maneuver is still impressive in an aerobatic routine today.

Katherine organized a six-month tour to demonstrate the airplane throughout China and Japan in 1917. She was among the first pilots in the Far East and certainly the first female pilot. In the same year, she set a record with a nine-hour-and-ten-minute nonstop flight from San Diego to San Francisco. In 1918 Katherine toured extensively; she set a number of Canadian distance and endurance records and also flew the first airmail in all of Canada.

Like other American airwomen, Katherine volunteered for combat pilot service in the war. She was, of course, also turned down, but was encouraged to make fundraising flights for the Red Cross and the Liberty Loan bond drives. These trips, especially one from Buffalo to Washington in 1917, were very successful and established Katherine as a pioneer in aerial public relations.

Katherine's persistent efforts to participate in the war effort finally resulted in a stint as an ambulance driver in France. This was certainly a curious deviation from an aviation career and may have been the origin of a debilitating illness she contracted, probably influenza or tuberculosis.

Constantly seeking roles equivalent to those of male pilots, Katherine repeatedly sought to become a pilot for the newly formalized U.S. Air Mail Service. In 1918 the Postmaster General finally approved her appointment. This important victory for gender equity was soon lost, however, as Katherine resigned from service after only two flights. Ill health may have contributed to this decision, for she never was an active pilot again. Marjorie, likewise, did not maintain an active aviation career. After the closing of the flightschool, there is little record of her actively flying. She pursued a career in mechanical drafting for the U.S. Navy Aeronautical Division (Underwood 1976; Oakes 1978; Parmalee 1972).

The Stinsons' contribution to aviation is amazingly full and well-rounded. They were not only performers, but innovators in performance. They were worldwide demonstrators of the practical utility of the airplane and pioneers in mail service, aviation education, and pilot training. It seems entirely fitting that the Army Aviation Branch would memorialize

Katherine as it did in 1988 by naming a new helicopter staging area in her honor (Kitchens 1989).

There were many other female American pilots in the years before 1920, though few as outstanding and active as Harriet Quimby, Ruth Law, or the Stinson sisters. Some simply flew but others established enviable records as exhibition flyers. Blanche Scott, who had already achieved some acclaim for driving a Willys Overland automobile from coast to coast, was the first female airshow pilot in the United States and probably the first woman to solo in this country. Known as "The Tomboy of the Air" from 1910 to 1916, Scott was often the featured pilot on airshow handbills. She apparently developed a reputation for daring as a result of low passes while inverted and flights under bridges. In any event, she impressed Glenn Curtiss, an early pilot, manufacturer, and general legend, who was not usually favorably disposed to female aviators; he trusted her as a test pilot.

Between 1912 and 1915 Alys McKey was one of the few show pilots on the Pacific Coast and in western Canada. Taking a back seat to no man, she received her share of symbolic honors, such as flying a demonstration for the Prince of Wales and the Duke of York. McKey was a tough customer indeed, continuing a flying career even after witnessing her husband's death in an airshow crash. McKey wanted to fly combat in World War I, but, like Ruth Law and Katherine Stinson, was turned down despite repeated inquiries.

Early flying was fairly dangerous, and both male and female pilots lost their lives as a result of error or equipment failure. Julia Clark had planned a sustained career as an airshow pilot, buying a Curtiss Pusher in which to practice her craft. Unfortunately she hit a tree at dusk in June of 1912. She was the first airwoman to die in a crash, preceding Harriet Quimby by just two weeks.

Barnstorming in the Twenties

"Barnstormers" was a label applied to a rough-and-tumble group of exhibition pilots who emerged after World War I. The name presumably reflects the fact that their landing strips were frequently pastures and their shelters were barns. Barnstorming grew dramatically, peaked, and declined all in the space of a decade. Although airplanes were increasingly common by 1918, the early demand for showmanship remained undiminished; indeed, because of public admiration of the real and imagined heroics of wartime fliers, it may actually have grown. The opportunities for meeting that demand, however, had increased dramatically. First, a large number of cheap, surplus airplanes were released from military service following the close of hostilities; these were mostly a standard trainer, the JN-4 "Jenny." Second, many military pilots and those who had taken flying lessons in the hope of becoming military pilots were unemployed and with little prospect of regular aviation jobs.

The result was a large number of one-person, one-airplane, itinerant businesses. Pilots followed the county fair circuits or just flew from town to town attempting to make a living selling rides or doing a few barrel rolls. Originally called "gypsies" because of their constant mobility, these adventurers trailed excitement in their wakes. In general, they were warmly welcomed; schools would even close so that children could see the miracle of the airplane. The barnstormers deserve a secure place in American aviation history, for they made flying a reality for most Americans, bringing it in tangible form to virtually every community in the country.

The effect of the barnstormers was not entirely positive, however. While there had been some important technical advances during World War I, airplanes were still not supremely reliable. The danger of mishap was compounded by two factors: the old Jennies were in poor condition when surplussed by the Army; and impecunious barnstormers were unlikely to invest heavily in regular maintenance. Moreover, demand for new acts escalated into stunt wars in which pilots would attempt increasingly bizarre and foolhardy maneuvers. The result, over time, was a decided decrease in the numbers of operable aircraft and living pilots.

Women as well as men were barnstorming pilots and many followed the classic professional path right into the ground. One of the most interesting examples is Bessie Coleman, the first black woman to earn a pilot's license and for that matter to pilot a plane in America. Coleman's desire to fly must have astonished her contemporaries. Such ambitions were far out of accord with contemporary expectations for members of her race. No flightschool in this country would accept her as a student, but like most airwomen, Coleman was ferociously persistent. She eventually trained in France, returning to the United States in 1921 as a fully qualified pilot.

Although she apparently had a long-term goal of opening a flight-school, Coleman joined the barnstorming brigade in 1922. Both her status as the only black female pilot and her considerable skill drew large crouds. She was successful in Chicago and throughout the South for four years. In 1926 her career came to an abrupt end while she was testing her plane prior to a major show in Orlando. Somehow she was thrown clear during a violent maneuver. Despite the years since Quimby's time, the use of seatbelts had not quite become standard practice. In any event, it would be years before prominent black women would fly again.

Of course, the possibility of mayhem helped to draw crowds, and barnstormers — some unconsciously and some shamelessly — exploited this fact. But thoughtful people soon saw that although dramatic deaths might draw people to shows, they did not instill public confidence in aviation. Mishaps jeopardized aviation's image as a miraculous boon to humankind; over the long term, death and destruction was bad business. Although reckless flying for the purpose of impressing others continues to this day, classic barnstorming could not survive the growing maturity of aviation. As we shall see in the next chapter, by the mid-twenties exhibition flying was frequently criticized in the newspapers, and a public safety campaign was underway.

Change, however, originated not with the press or external constituencies but with the aviation community itself. Exhibition flying continued, but it was somewhat more responsibly conducted. Organized commercial ventures, which had antecedents as far back as 1912, became more common. Multimember "flying circuses" replaced individual dilapidated Jennies flown by solitary, hard-bitten (and often hard-drinking) pilots.

Women were among the first to reassess barnstorming and hence to emerge as entrepreneurs in a new field. In 1921 Phoebe Fairgrave Omlie was a successful barnstormer, performing throughout the Midwest. Like that of many female airshow performers of the time, her forte was parachute jumping. She perfected a four-stage act: normal parachute descent, apparent failure of the parachute, free fall for a few seconds, and finally,

Mabel Cody, in an airshow exhibition, leaps from a motorcycle to a rope ladder suspended from an airplane, c. 1927. Smithsonian Institution Photo No. 82-3641.

close to the ground, opening of a second parachute and a safe landing. Omlie also established a record by jumping from the unprecedented altitude of 15,200 feet. Omlie, however, envisioned new heights; she founded her own airshow and took it on the road, the first woman to do so. Though successful, the organization soon diversified and evolved into a major fixed-base and charter operation, Mid-South Airways of Memphis, which was owned for many years by Omlie and her husband, Vernon.

Mabel Cody's experience was similar. An established barnstormer by 1924, Cody's preserve was leaping from a speeding car to a rope ladder dangling from a plane. Business sense competed with derring-do, however, and Cody soon established her own operation, "Mabel Cody's Woco-Pep Flying Circus." It turned out to be one of the more successful and spectacular such operations in the southeast. Although Cody continued to perform, once climbing from a powerful speedboat to a low-flying airplane, there was no doubt as to who was in charge. A surviving handbill announces C. Martin (a male) as a "pilot for Mabel Cody," and proclaims that Mabel Cody "flies with Woco-Pep, King of Motor Fuels." In an interesting bit of role reversal, the same handbill heralds the skills of a male parachute jumper named Barney (Smithsonian Institution photo 80-700).

From Professional Performers to Professional Practitioners

As classic barnstorming declined, the aviation community looked beyond exhibition flying for new professional challenges. A great surge in the practical dimension of aviation was on the horizon. The process of putting the airplane to work was fueled by innovation within the aviation community itself as well as by fairly explicit economic demand from other segments of the society. It should come as no surprise that women, already well established as pilots, were part of these developments as well.

The mid- to late twenties saw the more or less simultaneous emergence of at least nine practical uses of the airplane. To be sure, some of these had precursors a decade earlier and others were not destined to reach maturity until at least a decade later. But these aeronautical functions were clearly regarded as the wave of the future as barnstorming began to wind down: airmail, business flying, crop dusting and pest control, emergency/aerial ambulance service, forest-fire patrol, journalism and photography, map-making/public policy planning, passenger service, and prospecting/exploration. With astonishing rapidity, aviation's agenda had become comfortably full. The need for technocrats mushroomed accordingly.

During this period, interest in aviation performance by no means declined. Much as early individual barnstorming gave way to organized exhibition teams, so were the latter supplanted in the public imagination by an interest in aviation records for speed, altitude, endurance, or distance. Record setting, though part of aviation from the very beginning, came into its own only with the development of more powerful, more sophisticated, and more specialized aircraft. It differed from barnstorming or popular exhibitions in several respects. Instead of slapdash wild flying, it required the utmost in mechanical expertise. Maximum performance from engines and other equipment could be achieved only with great care, and, in most cases, only by investing a great deal of money. Record setting and racing were of great interest to aviation manufacturers, who saw them as opportunities to display both the performance capability and the safety of their machines. In addition, these activities were more closely linked with technological advances than earlier modes of performance. There is no doubt that the challenge of setting records was a direct stimulus to designers and engineers — complementing the demands of new practical applications — and encouraged systematic testing.

Aerial performance continued to charm the public, albeit in a new technocratic guise. Most ordinary citizens, of course, cared little about the extraordinary reliability of new radial engines or the details of oxygen systems; instead, they focused on personalities. They still wanted popular aviators, though perhaps a bit more well-groomed than the death-defying barnstormers. The result was the emergence of a class of professional aviation performers similar to the professional tennis players or golfers of today. They made their living by racing or going after records and they sought sponsors to support them. Unlike contemporary sports figures, however, they frequently did more practical piloting as well. Two days after an altitude record, one might fly a charter flight for business executives. A professional performer was likely to be a professional practitioner as well.

The number of female pilots who responded to these new opportunities was substantial, but hardly overwhelming. There were approximately five hundred female pilots during this period and various estimates suggest that about one hundred of them were unambiguously professional. In 1935 the magazine *National Aeronautics* conducted a survey of 142 female pilots and reported a similar level of professionalism. About 15 percent of this group earned a living in aviation (Studer 1935). A number of factors, however, portended an impact far greater than numbers alone would suggest.

Somewhat paradoxically, one of these factors was the aviation industry's use of women in a highly sexist fashion. In addition to the carnage caused by barnstorming, maturing aviation faced a second and related image problem. While ordinary people may have been favorably disposed toward aviation, few in the earliest years envisioned themselves as personal participants in flying. Indeed, some of the wonder and awe associated with "the winged gospel" suggests that such roles should be reserved for a select priesthood. Early pilots were accorded truly heroic status and many early public relations gambits sought to reinforce the image of near-superhuman abilities and virtuoso skills. This was all well and good for military aviation and for showmanship, but the idea of aviators as somehow larger than life was inconsistent with aviation's new domestic and practical agenda.

How could masses of citizens routinely patronize a wide variety of aviation services if aviation itself appeared inherently risky? If large numbers of Americans were to entrust their goods and their lives to airplanes, each flight could no longer be regarded as a death-defying adventure. Cult-figure status for the aviator had become counterproductive for business enterprise.

With a certain amount of cynical calculation, American entrepreneurs sought to tame the image of aviation by incorporating women into it. The underlying assumption was that women are inherently timid, unwilling to take risks, indecisive, uncoordinated, physically weak, and scatterbrained. If *they* can pilot airplanes — instead of only tall muscular men who were once champion athletes — aviation must be safe and piloting must be easy; perhaps commercial interests can be handled through aviation, and perhaps regular passenger travel really is possible (Corn 1979). Companies specifically hired female pilots and where necessary trained women to become pilots. These carefully prepared airwomen were then prominently displayed for public appreciation.

Judging by their comments, many airwomen knew perfectly well that they had been given a role as the frail sex in a larger promotional scheme. Though some found it demeaning and some wrote bitterly about it, many believed that the strategy would work and that all of aviation and the entire nation would benefit. Many made public comments along the lines of, "If I can pilot a plane, anyone can." Perhaps they agreed to accept a lesser role in order to promote the greater good.

In any event, it is clear that these developments created real professional opportunities for airwomen where few had existed before. This scenario was perhaps best reflected in the number of female sales representatives employed by airframe manufacturers and allied businesses. Sales representatives in the twenties and thirties, like their counterparts today, existed to demonstrate the performance capabilities of aircraft as well as to pitch creature comforts and other amenities. But given the newness of aviation and the fact that many customers were first-time buyers, demonstration of the ease of pilotage was perhaps an equally important function. It is this latter consideration that led leading companies to employ female sales representatives. Many of the best-known airwomen, including Louise McPhetridge Thaden, Elinor Smith, Ruth Nichols, and Amelia Earhart, held such positions at one time or another in their careers. So did lesser lights such as Blanche Noyes and Fay Gillis Wells. And of course there were those, like Charlotte Frye, who simply sold airplanes and neither sought nor achieved publicity or fame.

Thaden wrote insightfully about the role of the female sales representative. She did very well at Travel Air, a corporate ancestor of Beechcraft, and understood perfectly the relationship of gender to her job. She identified an interesting downside, however, to the male-female dynamics of aircraft sales circa 1930. Part of Thaden's role was to demonstrate planes to potential buyers and to ascertain their ability to handle a new type. Most customers accepted a woman serving in such a position of command, but a few men reacted poorly. When finally handed the controls of the demonstrator, these men felt the need to prove that no female pilot was superior to them. This usually took the form of extremely violent maneuvers. Far from showing powerful macho abilities, however, such antics threatened to exceed the limiting structural strength of the airplane. As pilot in command, and one who loved life, Thaden had no choice but to recover the controls. In some cases this involved wrenching them away. The resulting bruised male egos were not good for aircraft sales (Thaden 1938, 200–203). Female pilots no doubt helped convince America that flying was for all. Those who viewed themselves as among a select breed needing to demonstrate prowess, however, were apparently somewhat put off.

Most professional airwomen of this period labored in relative obscurity, but a disproportionately large number of them were attracted to racing and record setting. The glare of performance-generated publicity illuminated the careers of many, advertising the fact that a large cadre of professional airwomen existed. The same cultural norms that validated early female performers ensured that female record setters were newsworthy. Again, in a seemingly paradoxical way sexism may have generated opportunity. That is to say, the development of professional prestige for airwomen was, to an even greater degree than for their male counterparts, dependent upon performance as opposed to more directly practical work.

This is especially clear in the cases of two well-known airwomen, Elinor Smith and Ruth Nichols. These individuals were very different from one another: Smith came from a nontraditional background and parents who encouraged independent self-realization; Nichols emerged from the most genteel of circumstances and her life-long passion to fly was not at all compatible with parental hopes and dreams.

Smith was born in 1911 to a father who was a major New York vaudeville star (Tom Smith) and a mother who was a showbusiness singer. The Smiths were often on the road playing various towns. Family life was anything but traditional and there were no domestic role models to follow. On the contrary, independent-minded women were apparently common in the theater community, and Smith's parents were very supportive of her interest in airplanes. They felt that she should have every opportunity to develop her potential as she saw fit. Tom Smith was an early advocate of practical aviation; he was unusual among showbusiness people in that he actually hired a pilot to fly him from city to city to meet his engagements. He later took flying lessons himself and became a successful pilot. Smith was attracted to professional piloting early and was demonstrating the highly popular Waco biplanes by the time she was sixteen. Although stunts and exhibition flying were part of her agenda, at a very early age Smith looked to practical business to make a living (Smith 1981).

Smith quickly grasped that opportunities for professional advancement would be improved by public attention. So motivated — and probably also stimulated by frequent verbal jousting with her male counterparts — she undertook in 1928 a foolhardy stunt reminiscent of barnstorming in its heyday. Her flight under Manhattan bridges, discussed in Chapter 2, most certainly brought her publicity if not universal respect for good judgment. This, in turn, enabled her to find sponsors for an endurance flight.

By 1929, the Federation Aeronautique International (FAI) had established itself as the worldwide official recording body for all kinds of aviation records and had declared women a particular class of pilots. This meant that, for example, highest speeds and greatest altitudes achieved by female pilots would be published separately from those of men. Though many were uneasy about the implication of inferiority that such a separate designation might imply, there is no doubt that the existence of new official records attracted much attention. Smith and other female aviators understood this, as did airplane manufacturers, oil companies, and others interested in the advertising that accompanied sponsorship.

The sudden appearance of many new record categories led to an almost comical sequence of records broken and then almost immediately broken again. Although Stinson's 1917 San Diego–San Francisco flight had lasted longer, Viola Gentry set the "first" women's endurance record on December 20, 1928, by staying aloft for a total of eight hours. On January 2, 1929, Bobbi Trout flew for eleven hours. Smith took the record to thirteen hours at the end of January, but Trout recaptured it at seventeen hours in early February. Louise Thaden joined the mad rush with a twenty-two–hour flight in March. Smith ended this particular sequence in April with a flight of more than twenty-six hours. Smith, of course, as victor for the moment, was the primary beneficiary of this four-month-long drama (Brooks-Pazmany 1983, 28–31).

Smith followed this with other endurance efforts, upping the total to forty-two hours in a joint flight with Trout. She also held the women's altitude record for a time, as well as certain classes of speed records. Her attempts to regain the altitude record, after losing it to Ruth Nichols, attracted a great deal of attention; she simultaneously mounted a serious assault on the absolute altitude record, held by a man. In short, Smith was a self-conscious and aggressive performer and record setter. She also intended to make a living as a pilot and wanted to serve as a role model to women with similar aspirations. She regularly presented proposals for various record flights to sponsors and in the process secured remuneration for her aeronautical efforts. She thus became one of the better-known professional performers.

Smith, however, was not unfamiliar with the role of professional practitioner. On the contrary, she was closely associated with Giuseppe

Bellanca, a pioneer in airframe design. Many of her records were set in airplanes manufactured by the Bellanca firm and it is clear that in significant measure she was a test pilot and aeronautical researcher for that company. Bellanca is particularly noted for his development of large, practical airplanes capable of carrying substantial loads and for his experiments with engine supercharging. Smith was a valued associate. In addition, Smith held a number of straightforward piloting jobs throughout her active life, working for such firms as the Irvin Airchute Company and Fairchild Aircraft and Engine. In 1930 Smith was honored as one of the three best pilots in the United States by a group called the American Society for the Promotion of Aviation. Smith also broadened her professional career by becoming an aviation commentator for NBC radio and by writing columns and articles for a number of national publications.

An interesting footnote to Smith's professional career is her claim that George Palmer Putnam once sought to engage her services as pilot for Amelia Earhart. Putnam, a New York publisher, in some sense "managed" Earhart, arranging her various flights and appearances, publishing her books, and eventually marrying her. Smith had a very high opinion of Earhart's courage and determination, but a low one of her piloting skill and her level of attention to basics. Her contempt for Putnam, who she believed drove Earhart to unwise actions and even to her death in the Pacific, was extreme. Smith alleges that Putnam wanted to co-opt her so that she would be removed as a possible source of competition with Earhart; hiring her would presumably accomplish that. Smith also claims that Putnam wanted her to make some flights that could be attributed to the less-skilled Earhart (Smith 1981, 104). Whether such a dishonest contract was ever really offered we will never know.

Ruth Nichols (1901–1960) was from a prominent New York family. Her father held a seat on the New York Stock Exchange and, as a member of the elite East Coast "400," circulated in the highest society. Though Nichols was urged to excel, particularly in athletics, she was expected to conform to a very typical set of standards for a woman of her station. Her parents were appalled at her early desire to attend college and become a physician. Her perseverance was only partially successful: she did graduate from Vassar, but the medical profession eluded her. Moreover, under parental pressure Nichols interrupted her education so that she could emerge as a society debutante. Although she quickly rejected this lifestyle, her social schedule took her frequently to Miami. Boredom and geography thus led her to Rogers Airlines, which operated a flying boat service between New York and Miami; she quickly earned a seaplane pilot's rating. Nichols was a copilot on a record nonstop New York–Miami flight in 1928. The tremendous publicity that followed focused on her gender: headlines described Nichols as a "Daring Deb Flyer" and as "Lady Lindy." She was immediately hired by the Fairchild Aircraft and Engine Company in sales promotion. A record performance had led to a job (Nichols 1957).

Nichols, having come by her first job suddenly and without much planning, quickly became committed to an aviation career. She took a position with Aviation Country Clubs, a business enterprise that intended to establish a series of exclusive social and recreational organizations centering on flying, the new sport of kings. Her job was essentially cultivating prospects, flying out to wherever they might be. Although Nichols and others worked hard at this idea, and a few such clubs actually were established, no string of linked units ever developed in the United States. Nichols found work with Clarence Chamberlain, a prominent aviation figure who made a dramatic Atlantic crossing in 1927 — but a few days after Lindbergh. She aided in tests and demonstrations of some of his new airframe designs, ferried airplanes to remote sites, hauled passengers, and delivered overnight New York newspapers to midwestern cities.

Although thus reasonably well established as a professional practitioner, Nichols sought a greater challenge in performance and record setting. Declining family fortunes probably made cash prizes seem attractive as well. With Chamberlain's technical assistance and a borrowed Lockheed Vega, Nichols began a series of flights designed to culminate in the first solo crossing of the Atlantic by a woman. In 1930 and 1931, she set a women's record for an east-west, coast-to-coast flight, an altitude record, and a marked course women's speed record, all of which generated a great deal of publicity. In June of the latter year Nichols set out for Europe from New York, amid the shouts of a great crowd and with the honor of an initial military aircraft escort. Unfortunately, she crashed on landing at St. John's, Newfoundland, greatly damaging the plane and breaking several vertebrae.

Recovery was swift and Nichols promptly broke the women's non-stop distance record with a flight from Oakland to Louisville — a feat made all the more dramatic because she was still in a cast from her previous injuries. A crash on takeoff from Louisville, however, required complete rebuilding of the plane. When she immediately crashed the rebuilt version in New Jersey, Nichols's hopes for an Atlantic crossing — and for any other additional records — were dashed. The New Jersey accident was at least the fifth major aerial mishap that had befallen Nichols.

Nichols reverted to the role of a more conventional professional pilot, though in the Great Depression pickings were slim. She endured one more serious injury in a crash resulting from engine failure in an old Curtiss Condor she was co-piloting. Thereafter Nichols tried to link professional flying with personal humanitarian motivations, which she had apparently acquired from a Quaker aunt. In 1937, with war on the horizon, she made an unsuccessful attempt to organize an around-the-world goodwill flight under the auspices of something called the Emergency Peace Campaign. In 1939, she organized "Relief Wings," a volunteer aerial ambulance service to be mobilized in the event of a natural disaster or invasion. Although nearly everyone approved the concept, it may have been the product of overly enthusiastic airminded naivete. The most successful operation of this unit seems to have been an elaborate drill on Long Island. Though not dramatically successful, Relief Wings's functions were considered important enough to be formally absorbed by the Civil Air Patrol in World War II (Oakes 1985, 49–51).

Though very different from one another, both Smith and Nichols were thoroughgoing professional airwomen. Performance and record setting brought them into the public eye, and both understood that at least some of this attention was a function of their gender. Though in a sense uncomfortable with this situation, they consciously decided to use it to their own professional advantage — and that of other women as well. Frequent guests of various professional women's organizations, Smith and Nichols both hoped to serve as an inspiration for others.

Setting records and winning races was by no means a ticket to riches for the airwomen — indeed, obscurity was the eventual lot of many and one, Laura Ingalls, ended up in jail as a convicted spy. But many female pilots less prominent than Smith and Nichols followed their example of

Phoebe Omlie (left), participant in the 1936 presidential campaign. Also pictured is Stella Aikin. Arnold Collection (Negative No. 3650). Courtesy National Air and Space Museum.

converting public adulation into access to various kinds of professional opportunity.

One particularly interesting example involved traditional politics. Phoebe Omlie, the reformed barnstormer, established a fine reputation as a businesswoman and professional pilot. She enhanced that by remaining active — and victorious — in the cross-country racing scene. Most notably, she was the overall winner in the 1931 Transcontinental Handicap Air Derby, a race from Santa Monica to Cleveland, in which she bested some fifty male pilots. Omlie came to the attention of presidential campaign staffs in 1932. Then as now, candidates sought the endorsement of celebrities and tried to identify themselves with successful new technologies. The Democratic National Committee engaged Omlie to fly all over the United States campaigning on behalf of candidate Franklin Delano Roosevelt. Her trip covered over twenty thousand miles and she spoke to countless potential voters. While the effectiveness of this effort is difficult to assess precisely, it certainly was a trendy and attention-getting way to campaign.

The Roosevelt camp, in its success, was grateful. After inauguration, as time-honored tradition dictated, Omlie received an appropriate political appointment: Special Adviser for Air Intelligence to the National Advisory Committee for Aeronautics. This led to her directorship of the National Air Marking Program. Far from a traditional political sinecure, this position allowed Omlie to make a significant contribution to aerial navigation. She conceived and developed the idea of a series of large, rooftop markings that would identify locations to airplane pilots overhead. In order to appreciate the significance of this plan, one must remember that almost all flying in those days was "VFR," that is, visual flight rules. Pilots depended on a view of surface landmarks to determine their courses and locations. This is in contrast to "IFR," instrument flight rules, the standard for commercial aviation today, in which navigation depends on electronic devices and involves no visual contact with the ground whatsoever.

The imprecision of landmark navigation sometimes resulted in fliers getting lost, as it still does among light airplane and recreational pilots who necessarily continue to use VFR. In Omlie's time, when there were no VOR or LORAN navigation systems and no air traffic control, this was a serious problem. Omlie proposed painting the names of towns as well as elementary directional data on prominent rooftops, typically of warehouses, gymnasiums, hangars, or other large structures throughout the nation. Omlie hired a staff of five flying administrators: Nancy Harkness Love, Helen McCloskey, Blanche Noyes, Helen Richey, and Louise Thaden. The physical work of actually laying out and applying the markings fit well into the Depression-era Works Progress Administration format. An extensive system of yellow painted markers appeared throughout thirty states (Oakes 1985, 9, 40–42).

The marker program was an unqualified success; Omlie and her cohorts received well-deserved credit for both an innovative idea and effective execution. The program, however, was short-lived. As World War II loomed on the horizon, panicky civil defense officials feared that the markers might help invading enemy airplanes to find their targets, and so many were covered over. The resumption of the program following the war was halfhearted. The failure of the expected general aviation boom discouraged heavy investment and the development of modern radios made markers obsolete.

Airwomen displayed many variations on the professional theme. Blanche Noyes, before she became an air marker, was a corporate autogiro pilot for Standard Oil, occupying a role that has changed little — except for the types of aircraft involved — from the late twenties to the present day. Others used their professional flying skills to advance a second career. In the case of Fay Gillis Wells, it was journalism; early work with Curtiss Flying Service led to correspondent status for various aviation magazines. She also earned assignments with several major newspapers. Apparently flying allowed her to get to the scene of the action quickly, which was highly desirable for a reporter.

Gillis spent considerable time in the Soviet Union as both a newspaper person and a pilot. One of the few English-speaking aviation professionals in that country, she handled local arrangements for Wiley Post, a famous aviation pioneer who passed through Russia on a solo round-the-world flight in 1933. She was so effective in this role that Post invited Gillis to be his co-pilot on a flight from Los Angeles to Moscow in 1935. Gillis, after some soul-searching, declined in favor of a newspaper assignment in Ethiopia. Post then asked humorist and American folk hero Will Rogers to accompany him. Post and Rogers died in a crash at Point Barrow in one of the most well-known aviation mishaps of the thirties. Gillis, by a narrow chance, thus lived on to enjoy a distinguished newspaper career (Pateman 1985).

Without doubt the most dramatic single event involving a professional airwoman occurred in 1934. The key player was yet another of the air markers, Helen Richey. Richey's career was perhaps typical of the professional/record-setting syndrome. Apparently interested in a professional flying career from the beginning, Richey flew exhibitions and airshows in the Pittsburgh area. As early as 1930, she served as a copilot on a commercial airliner, albeit on a one-time, demonstration basis. She took up both record setting and racing, staying aloft for almost ten continuous days in a 1933 endurance flight and winning the major event at the Women's National Air Meet in 1934.

Successful racing and record setting not only had publicity value, but also demonstrated basic competence in pilotage. Undoubtedly both of these considerations occurred to the corporate officers of Central Airlines when they hired Richey as co-pilot in 1934. She prevailed over eight male applicants for the position. Although other women had served as professional pilots of large, transport aircraft, Richey was the first pilot for a regularly scheduled commercial airline. Central had just been awarded a Washington-Detroit airmail route and was vigorously competitive. The airline needed competent flight crews, but also publicity. Richey filled both needs, serving not only in the cockpit but on the lecture circuit as well.

As it turned out, however, the relationship between airline and pilot was neither long nor happy. Richey was not pleased at being diverted to public relations jobs — she only made about a dozen round trips in a ten-month period. More importantly, there was clear objection to Richey's gender on the part of Central's other pilots; indeed, they denied her membership in their union. In addition, the Civil Aeronautics Administration (CAA), then the government agency responsible for regulating American aviation, put considerable pressure on Central. Although Central was not specifically ordered to restrict Richey's flying, it was "suggested" that she be allowed to handle only daytime flights in clear weather. The genesis of this extraordinary intervention is unknown; it was formally justified on the basis that women possessed less physical strength. Although Richey's work seems to have been perfectly adequate and the airline seems to have been as supportive as possible under the circumstances, the arrangement came to an end in October 1935, after less than one year. Richey, in despair, resigned (May 1962, 136–138). From a personal standpoint, it had not been easy to establish this important first for U.S. airwomen. As we will see in the next chapter, female pilots were well enough established by this time that the Richey-Central matter became very controversial.

In addition to the air marking program, Richey's subsequent career included a return to racing. She served as Amelia Earhart's co-pilot in the 1936 Bendix, a major transcontinental race. More interesting, perhaps, is that Richey went to England in 1940 to join the Air Transport Auxiliary (ATA). That nation was already at war and had devised a scheme to use the flying skills of female pilots in noncombat roles. Though most of the pilots were, of course, British, highly qualified volunteers like Richey were welcomed with open arms. She served as a cargo and ferry pilot, at long last a truly professional role.

Other professional airwomen combined performance with work in varying degrees. Some, like Jean LaRene, were at best failures on the performance front. They simply enjoyed flying and relied on competence to get their jobs done. Edna Gardner Whyte flew a few races and contemplated a transatlantic hop, but was much better known as an aviation businesswoman. In the thirties Whyte repeatedly sought employment as an airline pilot. Her failure in this regard only led her to increased efforts in aviation education. She founded flightschools and remained president and chief pilot of a major operation until well into her eighties. Mae Haizlip took racing seriously; in 1931 she achieved second place in total prize money earned among all race pilots, male and female. Haizlip was also a test pilot for Spartan Aircraft and worked for other firms (Whyte and Cooper 1991; Buffington 1969; Oakes 1985, 38).

The Celebrities

No assessment of this group of professional airwomen would be complete, however, without some consideration of those who were at the high end of the performance scale. Not surprisingly, these were the individuals who were also the most famous; their contribution included drawing public attention to female aviators. To be sure, this was just as important in the thirties as it was in the days of Ruth Law, but it is interesting to note that even the most revered of the racers thought of themselves as professional airwomen and had something of a nonperformance professional life. Moreover, at least two of the three discussed here sought actively, if imperfectly, to further the professional ambitions of other women. Theirs is a dual contribution, image oriented as well as rooted in the economy of employment.

Louise Thaden, Amelia Earhart, and Jacqueline Cochran had in common a very high degree of motivation, a somewhat undisciplined curiosity, and a strong need for adventure. It was perhaps somewhat by accident that these qualities were expressed through careers in aviation.

Thaden (1906–1979) grew up in Bentonville, Arkansas, one of two daughters. Apparently her father had wanted boys and encouraged her in athletic, mechanical, and other typically masculine pursuits. Thaden was urged to go to college (even though she did not formally graduate from high school), but her career at the University of Arkansas never resulted in a degree. She began her studies at only fifteen years of age and partially completed curricula in journalism, physical education, and premedical studies.

Thaden took a year off from college to work for a coal company in Wichita, essentially as a sales representative. Wichita was the home of the Travel Air Company. The firm was just beginning to make a name for itself and Thaden, in part because of business connections between the coal company and Travel Air, "haunted the airplane factory and the flying field" (Thaden 1938, 52). Walter Beech, president and CEO, recognized the advantages of women in aviation and hired Thaden in 1927 as a West Coast sales representative. Of course, the company had to teach her to fly.

The love of airplanes and flying that Thaden had developed in Wichita only became stronger as her flying skills improved. She became a great success, not only as a pilot but also as a salesperson and manager. Beech foresaw even more public relations and advertising benefits if this competent pilot could set some records using Travel Air airplanes. In December of 1928 special fuel tanks were installed in a stock biplane and Thaden suffered through twenty-two hours of solo flight to establish a new women's endurance mark. Thaden made the newspapers in great style. Though she was careful not be completely swept away by her own publicity, she did enjoy fame and attention, as did Walter Beech and Travel Air.

In April of 1929 she set a women's altitude record with an ascent of over twenty thousand feet. A specially manufactured set of wings permitted a new measured-course speed record as well. Thus, for a brief period — many people were in the record-setting business — Thaden held three international records at the same time. In a very short time she had become one of the best-known female pilots in the country.

Thaden's rise to prominence paralleled a dramatic rise in public excitement over all aspects of aviation. Though airmindedness had been prominent at least since 1910, the late twenties were a time of veritable

Louise Thaden (left), after establishing an endurance record in a Travel Air biplane, 1929. Others unidentified. Smithsonian Institution Photo No. 83-2092.

frenzy. Lindbergh's 1927 flight from New York to Paris seemed to trigger a new wave of aviation accomplishments and corresponding public interest. Crowds at aviation events increased enormously and press coverage mushroomed. By 1929 promoters were very busy indeed, as they tried to capitalize on nationwide enthusiasm.

While pylon racing and other forms of closed-course competition became quite popular at this time, cross-country and transcontinental events seemed to have the greatest appeal. In 1929 there was a great Women's Air Derby, a race from Santa Monica to Cleveland. The race was to take several days and was open only to contestants flying solo. Given aviation's postbarnstorming maturity and both public and professional expectations, many safety precautions were in place. Parachutes were required and food and water had to be carried in case of a forced landing in a remote location. Excitement was intense and press coverage was enormous. Twenty pilots entered, including Louise Thaden flying a Travel Air airplane. The race was a daytime affair, with all contestants obliged to spend each night at a designated airport. Total elapsed time between the various points would determine the winner. The entire race lasted eight days.

Though this was, in fact, a legitimate race, interest extended well beyond considerations of actual speed and technical accomplishment. The public craved background and human interest; promoters contrived to oblige. Stops en route involved news conferences, banquets, and social events, all of which galvanized interest and contributed to the fatigue of the contestants. One contestant was killed and there were several crashes and other assorted mishaps. However, fifteen of the twenty entrants completed the course, a remarkable achievement given the nature of aircraft and navigation equipment at the time. Louise Thaden emerged victorious.

The Derby was the key event in the National Air Races, a huge and varied set of competitions for both men and women and clearly the premier annual event for the aviation community. As such, this race securely fixed women in the public mind as serious aviators. Though there were many races in which women pilots subsequently flew, no event ever compared to the 1929 Derby.

Thaden was at the height of her powers and the pinnacle of her popularity. Instead of capitalizing on the moment, however, she seemed to lose her momentum. Having married aeronautical engineer and aircraft designer Herbert Von Thaden shortly before the Derby, she decided to settle down. Family life in Pittsburgh proved unsatisfying despite continued involvement in aviation through her husband's aircraft manufacturing company. Though Thaden raised a child and attended upon her husband, her restlessness eventually took her out of the home. She took a position with the women's division of the Pennsylvania School of Aviation and another as public relations director for an organization called the Pittsburgh Aviation Industries Corporation. Thaden even competed in another transcontinental race. The 1931 Cross Country Derby was also flown from Santa Monica to Cleveland. This time her plane was a "Thaden T-4," a cumbersome monoplane designed, manufactured, and marketed by her husband. The T-4, destined quickly to go the way of the passenger pigeon and the dodo bird, left much to be desired and Thaden did not do well (Buffington 1967).

In 1932 Thaden was called on by promoters planning a major endurance flight. The development of techniques for in-flight refueling generated some new opportunities. Thaden's skills were suddenly in demand. She and Frances Marsalis went aloft on August 14 over Long Island. Though the two set a record of 196 continuous hours in the air, the flight is perhaps more notable for the extent to which public relations controlled it. Promoters wanted Thaden and Marsalis to remain in the air for an additional eight days in order to fly to Cleveland, accompanied by press planes, to open the National Air Races. The two declined this opportunity, but they did stage a phony attack of appendicitis in order to generate increased media attention.

Thaden again returned to family life and child rearing, savoring the few times she could fly. These were becoming increasingly rare because her husband had taken a position with an airline and there were no longer company planes available. Thaden fought boredom by working a stint for the Air Marking Program discussed earlier. But another dramatic opportunity was to come her way.

During the thirties, the Bendix Corporation began sponsoring an unlimited transcontinental air race. Careful organization and substantial prize money quickly made it the most prestigious operation of its type. The biggest names and the most aggressive manufacturers were always well represented. Despite pressure and protest, women were not initially admitted as competitors. The substantial success of airwomen, however, made it difficult to maintain that position and with some grace the policy was changed in 1935. The two female entrants in that year's Bendix did not fare well: Jacqueline Cochran was forced to withdraw en route and Amelia Earhart finished last. This was, however, no indicator of the future. Beechcraft corporation, formerly Travel Air, sensed opportunity. The firm had not forgotten Louise Thaden's contributions to the company's good public image. Once again Thaden was called out of her racing retirement. For the 1936 Bendix race, the company provided one of its new cabin biplanes with retractable landing gear and Thaden gladly accepted the challenge. Fellow air marker Blanche Noyes served as co-pilot on this New York–Los Angeles run. They won in a time of just under fifteen hours.

With second place taken by Laura Ingalls flying solo, the 1936 Bendix signalled that women pilots had truly arrived (Dwiggins 1965). In the 1929 Air Derby, women demonstrated that they could conduct a grueling cross-country race; in the 1936 Bendix, they demonstrated that they could beat the very best of the male pilots.

This dramatic victory turned out to be Thaden's last hurrah. She accepted a second term as sales representative for Beechcraft and enjoyed a solid professional career for a few years. She set a few minor speed records in 1937 (for example, from Detroit to Akron in forty-one minutes) but she never raced again. Thaden quickly disappeared from the aviation scene, devoting herself to family concerns.

Thaden's writings, which will be discussed in greater detail in the next chapter, show her to be a very complex individual. That she was an adventurous person in need of diverse stimuli and frequent change is evident even from the brief description above. In addition, she chafed at the various restrictions and prejudices that kept her from full self-realization as an aviator. Moreover, she was able to view these from an abstract viewpoint and grasp the difficulties faced by all women in this culture. Though she did not write for women's magazines or accept speaking engagements with women's professional groups, her words betray a sophisticated feminism. On the other hand, Thaden was clearly committed to the values of the traditional family, including at least in some respects obedience to her husband. There is no doubt that this occasioned a good deal of guilt and internal conflict. Judging by her on-again–off-again career and her failure to aggressively articulate her views, she never completely resolved it.

Amelia Earhart (1897–1937) was the best-known female aviator that ever lived. This is due to three factors: the objective impressiveness of her aeronautical accomplishments, the skill with which her public relations campaign was orchestrated, and the great drama and mystery surrounding her disappearance over the Pacific. Indeed, she is the only airwoman known to the vast majority of Americans today.

Like many of the airwomen, Earhart had a somewhat unconventional childhood. Because her mother was from a prominent Atchison, Kansas family, Earhart began with many of the benefits of social position and wealth. The promise of typical midwestern gentility, however, was not

Amelia Earhart and her instructor, Neta Snook, c. 1921. Smithsonian Institution Photo No. 82-8666.

to be realized. The nuclear family was radically out of accord with what might have been expected. Earhart's father was an attorney, apparently of some initial promise, for the Rock Island Railroad. He was apparently an alcoholic, however, and this led to downward social mobility as well as geographical mobility motivated by the search for new railroad jobs. Complete disaster was forestalled by a modest inheritance and Earhart enjoyed some amenities, such as fashionable private schools. Nonetheless, her social marginality was strong enough to generate a set of attitudes that was substantively unusual, although not really radical or revolutionary. Her view of the future was rather unfocused and she seemed indecisive in personal relationships and professional plans (Morrissey 1963). She contemplated a number of futures, including business and medicine, but nothing came to fruition as she remained dependent on family well past adolescence (Lovell 1989, 11–52).

At the age of twenty-three, Earhart began to develop an interest in flying; apparently she was stimulated by the high level of aviation activity in southern California, where the family was located in 1920. Still under the influence of a father who disapproved of her spending long hours with a male instructor, Earhart learned to fly under the guidance of Neta Snook Southern, one of the earliest of female instructors (Southern 1974). Earhart did some incidental flying in California and then moved to Boston with her mother and sister in 1924, still apparently at loose ends. In 1925, despite a complete lack of appropriate training, she took a position as a social worker at a settlement house. Though Earhart did not fly much during these years, she did attract some publicity as a promoter of aviation — particularly women's aviation — and she was director of a firm that operated an airport and a flightschool.

Against this lifestyle, characterized by lack of purpose, lack of vision, dependency, and relative obscurity, there appeared George Palmer Putnam. Putnam controlled a large family firm that had been a major New York publishing enterprise for many decades. In the mid-twenties, the firm concentrated on popular nonfiction spanning a vast range of topics. Putnam was not only an executive; he counted among his tasks getting out on the street and hustling profitable book titles. Putnam went far beyond the usual solicitation of authors; he truly became involved. He personally organized and lead expeditions to remote and unexplored areas, observed local exotica, and published the resulting accounts of the explorations. Even attempting a record flight of his own (it failed) was not too extreme a method for Putnam to generate markets for books. Nor, reportedly, was playing fast and loose with facts. In 1929 a series of sensational and publicity-generating threats against the firm of G. P. Putnam's Sons attended the publication of an anti-Fascist book by an Italian author. The fact that the threats were phony, apparently written by Putnam himself, did not prevent book sales from skyrocketing (Lovell 1989, 158).

In the months immediately following Lindbergh's flight from New York to Paris, aviation was certainly a hot topic for the publishing industry. Putnam, accordingly, could be found checking out New York's various airports in early 1928. He discovered that several women, including Ruth Nichols, were interested in being the first to cross the Atlantic. By way of competition, certain sponsors contemplated a flight in which "the right sort of girl" — apparently someone acceptable to members of English society — would be flown across as a passenger. Putnam joined the project and set out to find the appropriate person. Through a chain of business contacts and their friends, he eventually came to Earhart. She was judged to have the proper breeding. Moreover, her general physical resemblance to Charles Lindbergh (tall, lanky, and blond) was immediately recognized for its publicity value; a Putnam associate had coined the "Lady Lindy" tag before the deal was finalized. In any event, Earhart was chosen without any regard to piloting abilities.

On June 18, 1928, Wilmer Stultz and Lou Gordon flew from Trepassy Bay, Newfoundland to Burry Port, Wales in a Fokker Tri-Motor fitted as a seaplane. Amelia Earhart, as passenger, was the first woman to cross the Atlantic by air. Careful orchestration of the media ensured that she became famous overnight. Earhart was put on the lecture circuit for a few years and otherwise became an object of the Putnam publicity machine, marrying George Palmer Putnam in 1931. She was in the news for various feats including deep-sea diving and air racing, but dramatic success as a pilot eluded her. Continuing publicity demanded exposure, however, and so she undertook the setting of rather obscure records — for example, in 1931 she set the record for altitude in an autogiro of 18,451 feet.

A more dramatic accomplishment was deemed necessary and a solo transatlantic flight was planned for 1932. Again, the attention-getting element of competition was present, for both Elinor Smith and Ruth Nichols were contemplating such a jaunt. Putnam, ever concerned about publicity, set the date of the attempt for May 20, the fifth anniversary of Lindbergh's effort. Earhart flew a Lockheed Vega from Harbor Grace, Newfoundland to Londonderry, becoming the first woman to pilot a plane across the Atlantic and only the second person to do it solo. The feat was indeed remarkable, demonstrating once again that airwomen, given the proper equipment and financial backing, were capable of the same sort of performance as their male counterparts.

Earhart continued in this vein with a solo flight from coast to coast later in 1932, the first such by a woman, and a flight from Honolulu to Oakland in 1935, the first solo by any person. A solo flight from Mexico City to Newark, also in 1935, emphasized not only that women were perfectly well suited to long-distance pilotage, but also how famous she had become. A crowd estimated at up to ten thousand met her at her destination. Putnam, characteristically, had arranged for hundreds of autographed envelopes, with special Mexican postage stamps, to be carried on the flight. Subsequent sales to collectors were brisk.

Success, however, caused its own problems. In a sense, Earhart and Putnam were running out of worlds to conquer by mid-decade. Earhart's performance in races had always been lackluster and there were few additional "first" flights that the public would find exciting. A round-the-world venture seemed about the only available alternative. Such a flight would not be a first. Wiley Post, for example, had accomplished it twice, but his flight path had been entirely in the Northern Hemisphere. Earhart's planned route was generally equatorial and as such involved greater distances and much more difficult navigational problems. In short, even as late as 1937 such a trip would be regarded as significant if not monumental.

In a very real way, however, planning and financing the venture was more challenging than the actual effort of piloting and navigating the airplane. Clearances and permissions took time and patience to acquire, and the twin-engine Lockheed Electra, with its several custom modifications, was costly. Putnam's business and political connections were equal to the tasks of cutting through bureaucracy and finding the money. Putnam, a friend of the president of Purdue University, suggested the airplane itself should be provided by the school as a "flying laboratory." Though it is dubious that the Electra was ever used for the systematic collection of scientifically relevant data, Purdue University Research Foundation was the vehicle for the accumulation of funds for its purchase. The connection with Purdue had been in place since 1935 when Earhart began a period as consultant to the Women's Careers Department.

The first attempt began on March 17, 1937. The flight was to be flown from east to west, the first leg from Oakland to Honolulu. The second leg, from Honolulu to Howland Island, never got off the ground as Earhart ground-looped on takeoff and severely damaged the plane. It was returned by ship to California and repaired. A second attempt was begun on May 21, this time to the east. The starting date was to become highly controversial: many thought that additional checks remained to be done on the airplane, but Putnam had arranged for gala celebrations to occur on July 4, the planned completion date for the flight, and delay would have disrupted the schedule. If there is one fact generally known about any American airwoman it is that Earhart, along with navigator Fred Noonan, disappeared over the Pacific en route from Lae, New Guinea to

Howland Island. Mystery persists to this day; inquiries and excavations to discover what really happened still draw investors and heavy newspaper coverage.

The Earhart phenomenon is, to say the least, a strange and ambiguous one. There is no question that she possessed great courage and was willing to undertake huge challenges. On the other hand, she does not seem to have been particularly skilled as a pilot and many observers found her to be careless. Even a very sympathetic biographer noted widespread concern on this account (Lovell 1989, 250–259). Her crash on takeoff from Honolulu was just one of a long series of mishaps that, though blamed on mechanical malfunctions, were quite possibly the result of limited skill or inattention.

Moreover, despite a long record of activities that were unusual for a woman — and even some record of organized feminism — Earhart showed a consistent indecisiveness and dependency. To be sure, she retained her family name in marrying Putnam, highly unusual in that era. She belonged, at least nominally, to the National Women's Party, a very forward-looking organization that anticipated some of the issues that would become prominent in the sixties. She did not follow standard expectations in matters of dress and fashion. Yet she needed her father's support and permission, at the age of twenty-three, to take flying lessons. She was still essentially living with her family at thirty-one, when the Atlantic passenger trip was offered to her. And, most importantly, virtually her entire career as a pilot was managed by George Palmer Putnam for the benefit, critics would say, of his business enterprises. One might even ask whether the expressions of feminism were themselves contrived: Earhart's radical, egalitarian views about her marriage to Putnam certainly made attention-getting copy — and they were prominently published by George Palmer Putnam after her death. Despite the folk-hero status Earhart has attained, it is impossible to regard her as an entirely admirable personality. As an exemplar in the quest for full citizenship for women, she had feet of clay; we will consider this further in later chapters.

Jacqueline Cochran (c. 1908–1980) had two things in common with Earhart. First, she became one of the most prominent female pilots in the United States; second, she accomplished this at least in part with the help of a wealthy and influential man whom she eventually married. Beyond these two dimensions, however, the two were radically different kinds of individuals though they did, presumably, share some degree of friendship.

Cochran's youth was truly extraordinary. She was born in extreme poverty in rural Florida and never knew the exact date of her birth, let alone the identity of her parents. Her roots were so anonymous that when it became necessary for her to be identified by a surname, she arbitrarily choose one out of the Pensacola telephone book (Cochran and Brinley 1987, 49). Almost completely uneducated, Cochran rose to international fame through an incredible exercise of will.

Cochran had only two years of what passed for formal schooling. Her early years were spent working in lumber and textile mills in Florida and Georgia, enduring the tough conditions that child labor laws had yet to banish from the American scene. She was always ambitious and adventurous, constantly planning a dramatic future. At ten or twelve she was an inspection room supervisor in a weaving mill and shortly after she found a job in a Columbus, Georgia beauty parlor. Aggressive self-promotion soon led to better positions, both as a beautician and as a manager, in Montgomery and other cities throughout the South. Eventually, Cochran found her way to New York and went to work for Antoine's in Saks Fifth Avenue. As hairdresser to the nation's elite, her lifestyle changed dramatically. She appears to have used every possible contact to advance her career. She managed to penetrate the social scene of her clients and attracted the attention of financier Floyd Odlum, who would later become her husband.

In 1932, while on vacation from Antoine's, Cochran learned to fly. With sponsorship from Odlum and others, she flew a great deal the following year and earned various ratings including a commercial pilot's license. Ambition and insistence soon led her to the most powerful figures in American aviation. Cochran entered the highly publicized Mac-Robertson London-to-Australia race in 1934. Through various sponsors, she had obtained a Granville Brothers Gee Bee racer for this event — no mean feat since this was one of the premier racing plane manufacturers of the thirties. She was immediately a credible contestant. While mechanical problems and a longer-than-expected landing roll kept Cochran from proceeding farther than Bucharest, she had only begun her racing career. In 1935 she entered the Bendix, America's most prestigious

Jacqueline Cochran, after flying the Seversky Pursuit to victory in the 1938 Bendix Trophy Race. Also pictured are Vincent Bendix (left) and Alexander de Seversky. Smithsonian Institution Photo No. 84-14781.

transcontinental air race, with an ultramodern Northrop Gamma. Although she again withdrew because of mechanical problems, the effect of the two attempts was to clearly announce her as a major personality in aviation.

Cochran had invested in several hairstyling salons and in 1934 began, with Odlum's help, a cosmetics manufacturing business. This was to compete with aviation for her attention as Jacqueline Cochran Cosmetics became one of the largest such firms in the United States. Cochran eventually achieved success in the major race field, taking third place in the 1937 Bendix. She flew a Beechcraft similar to Louise Thaden's victorious machine of a year earlier and with it she also set an absolute women's speed record the same year.

At about this time, Alexander de Seversky, a Russian-born aircraft designer, was attempting to sell his radical new fighter plane to skeptical Army officers. He engaged Cochran — perhaps in yet another example of "if a woman can fly it, anyone can" — to aid in demonstrating the new machine's potential. She set another women's speed record in the plane and established the absolute record for a flight from New York to Miami.

Seversky's fighter eventually evolved into the Republic P-47 Thunderbolt, surely one of the most significant aerial weapons of World War II.

Cochran's connection with Seversky continued and she flew a similar plane, the P-35, in the 1938 Bendix. This was the pinnacle of her racing career: she won hands down, defeating all of her male competitors. With the second female victory in this race, only the most hidebound could deny that women were competent aviators. In 1939, Cochran established yet another women's speed record and a women's altitude record. Her attempt in that year to become the first two-time winner of the Bendix was frustrated by deteriorating weather conditions. As a finale, in 1940 she established with the Seversky airplane an absolute world speed record of 332 miles per hour over a 2,000-kilometer course. These various accomplishments established Cochran as the foremost airwoman in the United States. This status, coupled with husband Floyd Odlum's membership in the nation's financial elite, was to permit her to play a key role in the first U.S. experience with women flying in military service (Cochran 1954, 168).

Military Airwomen

World War II was an established fact of European life by 1940 and most observers well understood that — in part because of the airplane — it was to be a new and different kind of conflict. To be sure, airplanes made an appearance in World War I. The exploits of pilots between 1914 and 1918 were certainly colorful and often spectacular, providing much grist for journalistic and literary mills. As we have seen, the wartime experience greatly affected American public interest in aviation and boosted its popularity. However, from a strictly military standpoint the airplane was of limited importance in the 1914–1918 conflict.

World War II, however, was vastly different. The expansionist powers, Germany in Europe and Japan in Asia, understood that aircraft had developed into sophisticated offensive weapons systems. Their attack strategies were unprecedented. Germany's rapid conquests relied on both fighter support and the effectiveness of divebombing. The significance of European air warfare was grimly established by the long series of air engagements in 1941 known as the Battle of Britain. Japan was similarly effective, particularly with naval aviation, in its expansion throughout Asia. The aerial attack on Pearl Harbor underscored the fact that traditional concepts of warfare were no longer adequate.

Aerial attack, the victimized nations quickly discerned, demanded aerial response. Pursuit of the air war, and thus avoidance of quick and total defeat, required both brilliant initiative and tremendous investment. One of the most basic problems was fielding an adequate number of combat pilots. Effective and rapid training was the key ingredient and both the United States and Great Britain developed very successful instructional programs without which history might have been quite different. Other approaches to the "manpower" problem were tried as well. At least three nations, Great Britain, the U.S.S.R., and the United States, systematically used female pilots in the war effort.

In Great Britain, there was a concerted effort to employ all pilots, even those who were considered unsuited for combat. Older men or those with minor physical disabilities were organized into the Air Transport Auxiliary (ATA). As the name implies, this unit had the responsibility of moving goods or personnel to appropriate locations in the British Isles and of ferrying combat aircraft between factories or repair depots and active bases.

By the late 1930s, women in Britain had a piloting tradition similar to that of their American counterparts. Barnstorming and exhibition flying, for example, were by no means only a male preserve (Gower 1938). Several notable British women record setters even managed an American dimension to their careers, among them Amy Johnson (C. Smith 1967), Lady Mary Heath (Heath and Murray 1929), and Beryl Markham (Lovell 1987). Hence, it is not surprising, with invasion looming on the horizon and national survival really in doubt, that in 1941 the ATA incorporated female pilots into the operation on an equal footing with men.

The ATA numbered about eight hundred pilots and at any given time about one hundred of these were women. The women's record was in all respects comparable or superior to the men's. Indeed, women's fatality rate was lower than their male counterparts' even though the duties assigned were the same (King 1956, 177). This rate did not approach zero, however, and some fifteen "ATA girls" lost their lives in the line of duty. Women continued to fly throughout the war as full members of the military with equivalent pay and benefits. They shared in such duties as delivering combat planes across the Channel following D-Day and in ferrying the early Gloster jets.

On one occasion, Opal Anderson, an American ATA volunteer, was assigned to ferry a captured German bomber. An escaping British prisoner was able to steal the Junkers JU 88 from a German airfield and fly it undamaged back to England. This was not only an inspiring act of courage, but also a brilliant intelligence coup, for detailed technical information on this important weapon had not before been available to the Allied powers. The British High Command's great trust in the ATA girls shows in the ease with which they assigned Anderson the task of flying the prize to the Farnsborough Experimental Station (Granger 1991, 18).

The Soviets, whose need for pilots was equally desperate, looked at women somewhat differently. Perhaps motivated by the egalitarian features of a socialist ideology, they took a more direct approach. Half a century before the debate about combat status for American military airwomen, the Soviets established at least three women's combat squadrons: one fighter, one light bomber, and one night bomber, the latter equipped with obsolete biplanes. Moreover, not only were these squadrons combat-ready, but they regularly flew the toughest missions. Some of the best female fighter pilots were transferred to crack units at the front and flew in the same formations and battles as their male counterparts. One, Lily Litvak, reportedly shot down at least twelve German planes (Myles 1981).

In the United States, deteriorating world conditions were evident by the late thirties, but reaction to them was decidedly restrained. Isolationist sentiment remained strong, especially in Congress. The threat seemed remote and preparations for war were fitful and unenthusiastic at best. An early attempt to develop a cadre of pilots, the Civilian Pilot Training Program, was formally justified as part of a general attempt to promote "airmindedness" — to stimulate the aviation economy and better prepare the nation for the coming air age. To the less complacent, however, the CPTP was viewed as a source of potential military pilots. This was certainly true of Civil Aeronautics Administration administrator

Robert H. Hinckley, who conceived the CPTP in 1938 and convinced the President to establish it.

This program was based at colleges and universities. These educational institutions provided instruction in aerodynamics, meteorology, and history while operators at local airports were contracted to provide the actual flight training. The program was very popular, with over 215,000 students enrolled by 1944, the year in which operations ceased.

Women were a regular part of the early CPTP; at most colleges they were enrolled in a proportion of about one woman to every ten men. In addition, several women's colleges supported all-female units. Over the previous thirty years women had carved out a place for themselves in American aviation and it appeared that this new public program was about to recognize and confirm it. Even the first steps in the militarization of the CPTP — a reaction to the deepening crises in Europe and Asia that even entrenched isolationism could not prevent — seemed to involve women as equal players. For a brief time in 1941, all participants were asked to sign nonbinding agreements to serve in the military in the event of war and to indicate which branch of service they preferred. When the nonbinding agreement became a formal pledge of eventual enlistment as a pilot-officer in the U.S. armed services, however, women's circumstances suddenly changed. Since the eventuality for which CPTP was now preparing was not available to women, they were banished from the CPTP altogether (Douglas 1991, 6–7).

This decision was controversial, even drawing a public objection from first lady Eleanor Roosevelt. Hinckley felt obliged to write to many protesting women in a fashion that seems somewhat conciliatory — and that certainly anticipated the flying roles that military airwomen eventually achieved in World War II. "If or when the time comes when trained girls are needed in non-combat work to release men for active duty, that will be a different situation" (Planck 1942, 150). Hinckley's words were apparently sincere: the CAA indeed argued that women could appropriately serve in the noncombat role of flight instructor, thereby freeing male instructors for combat roles in the event of hostilities. There were aggressive plans for the preparation of several hundred female instructors, but funds were never authorized. Only about fifty women, some products of the early CPTP days, served as civilian flight instructors throughout the war.

It is clear that from an early date there was serious thought of a military role for airwomen. However, no consensus developed on how or when to use them; indifference if not opposition remained common. While a military program for airwomen ultimately emerged, it took a great deal of promoting. Moreover, since there was no common conceptual ground, diverse planning and lobbying efforts were often at cross purposes. Though the program's foundations were thus a bit shaky, women did regularly and successfully fly military airplanes in World War II. The architects of this effort were many, including key generals, but the chief advocates were two airwomen, Nancy Harkness Love and Jacqueline Cochran. Both, as it turned out, favored the British rather than the Soviet model.

Cochran appears to have been the first to suggest an organized military role for American airwomen. By 1939 she was not only famous for her brilliant aeronautical accomplishments but also well connected in Washington. Husband Floyd Odlum was prominent in the business and social worlds and carefully maintained political contacts as well. Though Cochran and Odlum were to become staunch political conservatives, at the time they were friends of the Roosevelts and contributors to Democratic electoral causes. Cochran could thus write Eleanor Roosevelt with great credibility on the topic of women in aviation. Anticipating eventual U.S. entry into a major conflict, she suggested the formation of a noncombat group not unlike the British ATA; she justified it, as others had done, on the ground of releasing male pilots for combat. Though there was no immediate result, a seed of interest had been planted at the White House. Cochran began to promote the idea with others and by early 1941 was discussing it with major military figures like General "Hap" Arnold (Cochran 1954).

Others also foresaw great need for military pilots and likewise thought of airwomen. Indeed, it was Colonel Robert Olds, a member of Arnold's staff, who in early 1940 asked Nancy Harkness Love to compile a list of all women holding commercial pilot ratings. Olds was concerned with the potentially massive task of ferrying overseas-bound military airplanes from manufacturing plants to debarkation points. Love was one of the few women who had experience in exactly that role.

Love was a successful but not greatly prominent airwoman of the thirties. Though from a background of wealth and social standing, Love worked as an aviation professional with the Air Marking Program organized by Phoebe Omlie; as a test pilot with a manufacturing company; and with a sales firm, Inter-City Airlines, organized with her husband, Robert Love. Among her many prewar, civilian assignments was the ferrying of American manufactured warplanes to Canada for transshipment to beleaguered France. The resultant contacts with the U.S. Army Air Corps' Air Ferrying Command (later Air Transport Command Ferrying Division) generated the request from Olds. Love compiled the list, but there was again no immediate result.

Though there is no doubt of Love's abilities both as a pilot and as an administrator, subsequent contacts, connections, and circumstance ultimately set the stage for a second event leading to her involvement with military airwomen. Robert Love, a reserve officer, had been appointed Deputy Chief of Staff with the Air Transport Command in 1941 and Nancy Harkness Love, probably not entirely coincidentally, got a civilian job with the Ferrying Division of that command — the very organization whose interest she had attracted earlier.

In the meantime, connections and contacts likewise favored Cochran. In the years prior to U.S. involvement in the war, British Ferry Command, parent organization to the Air Transport Auxiliary, maintained an active search for volunteers in the United States. Cochran, apparently through General Arnold, got in touch with recruitment officers in early 1941. Her plan involved one of ATA's heavy responsibilities, the movement of American-manufactured aircraft from Canada to Britain. Cochran wanted personally to ferry a bomber across the Atlantic. This suggestion was very controversial because women had not yet been involved with military aircraft. Many saw tremendous publicity benefits, but others stressed the negative impact on morale if the plane were to be shot down en route. Apparently out of traditional chauvinism, male ATA pilots objected strenuously, even after Cochran handily passed all flight tests and a physical examination. Concern and resistance were swept aside, however, when Floyd Odlum — who, it must be remembered, was a very wealthy international financier — contacted his friend Lord Beaverbrook. Beaverbrook was prominent in British politics and at the time was Minister of Procurement. In June Cochran flew a twin-engine Lockheed Hudson to Prestwick, albeit with the imperative to delegate certain takeoff and landing responsibilities to a co-pilot. While multi-engine aircraft flights across the Atlantic had become rather routine, this was once again a dramatic demonstration of women's ability to handle the biggest challenges that aviation could present. It did not go unnoticed.

This event had two important consequences. First, Pauline Gower, the commander of British female military pilots, formally asked Cochran to recruit American airwomen as ATA volunteers. Second, upon her return to the United States Cochran was instructed by Eleanor Roosevelt to go forward with plans for a female pilots' program in the U.S. defense effort. Cochran was appointed to work with Arnold and newly promoted General Olds, officers who had already indicated a strong interest in female military pilots. Cochran submitted a formal plan that was indeed grand in scope. It envisioned a complete female pilots' division, to be self-regenerating over time due to an internal training program. Cochran overplayed her hand; Olds and Arnold would not agree to such a major organizational addition. Despite this apparent setback, the idea of military airwomen was now established. Arnold was committed along these lines when war eventually drew near and, as a result of White House pressure, he was bound to include Cochran. Cochran probably understood this, and she could afford to resign in apparent pique (Douglas 1991, 27–33).

Cochran was thus free by late 1941 to recruit female pilots from the United States for the British ATA. An abundance of volunteers enabled her to deliver twenty-four excellent pilots, among them Helen Richey. They offered superb service, many remaining with the British much longer than the eighteen-month initial commitment. Cochran accompanied her charges to England to await events, confident that her plans for an American organization would soon come to fruition.

In her absence, however, other developments proceeded apace. With the United States now formally at war, attitudes changed. As the severity of the crisis became increasingly clear, even hardened conservatives began to back away from the traditional U.S. prohibition of women in the military. Women, they saw, could alleviate a critical shortage of "manpower" in many noncombat roles. The Women's Army Auxiliary Corps (WAAC) was created and Oveta Culp Hobby was appointed to direct it. Though this had no immediate implications for the employment of female pilots, it demonstrated that cultural norms were shifting in a favorable way.

Changes were also occurring in Washington. Colonel (later General) William Tunner assumed command of the Ferrying Division. This officer had no previous experience with female pilots, but he was promptly in urgent need of people who could fly airplanes. Since his immediate subordinate was Robert Love, he knew not only of Nancy Harkness Love's successful piloting career, but also of her early contacts with Olds. Love's plan for the employment of female ferry pilots had been dormant for nearly two years, but Tunner, duly impressed, moved quickly to the implementation stage. Love was transferred to Washington to put the finishing touches on her ideas. Her straightforward scheme involved the use of the most experienced female pilots available. They were to be hired as civilians since female pilots could not be commissioned under existing legislation; militarization was planned for later. Because Love wanted to minimize the chances of individual failure, technical and skill requirements for the female pilots were more stringent than for their male counterparts. In order to defuse a potential source of criticism, their pay was to be less. Nonetheless, the Women's Auxiliary Ferrying Squadron (WAFS) was established, with Love as Director, on September 10, 1942 (Verges 1991, 44–61).

To many, it seemed that closure in the question of military airwomen had at last been achieved. Women's accomplishments as aviators were to be recognized and their service to the nation was to be gratefully accepted. This was immensely gratifying both for the airwomen and for the nation in need. For General Arnold, however, there remained the question of what to do about Cochran. The WAFS had been established during her absence and, she would argue, in violation of his pledge that she would occupy the preeminent role in women's military aviation. Arnold was fully aware of Cochran's aggressiveness, to say nothing of her and her husband's Washington connections. Cochran returned to the United States within days of the establishment of the WAFS, and Arnold's responded in a way that can only be regarded as politically expedient: he approved the creation of a second women's paramilitary pilots program. Accordingly, Cochran became Director of the Women's Flying Training Detachment (WFTD). Love's organization was service oriented, while Cochran's provided for the continuous training of women so there would be a permanent supply. Like the WAFS, members of the WFTD were civilians.

The WAFS received a deluge of applications and Love handpicked an original cadre of twenty-seven. In October of 1942, following various testing and orientation operations, they began delivering liaison and trainer aircraft. The planes were small and simple. This assignment was a deliberately cautious one, motivated by lack of knowledge about the new pilot corps and by traditional doubts held by many officers about women's capabilities. By December, however, the WAFS delivery rate of 100 percent had impressed some of the more progressive military brass. Operations expanded from the original base at Wilmington, Delaware, to several other sites and the WAFS were assigned to more advanced airplanes.

Cochran's WFTD training operation had set up shop in October and began graduating women (called "Woofteds") in April 1943. They were assigned to the WAFS, thus supporting the expansion of that organization. By the significant date of August 1943, a total of 216 WFTDs had qualified for ferrying assignments (Douglas 1991, 44–56).

Requirements for entry into the WFTD were far less stringent than those for the WAFS. The former was a training operation whereas the latter was a program to effectively employ women who already possessed a high level of skill and experience. WFTD applicants originally were required to have two hundred hours of flying time (as opposed to five hundred for the WAFS), but this total was progressively reduced to thirty-five by the program's end. The program was run in a fashion parallel to training for men, involving some 210 hours of actual flight training as well as other appropriate instruction. Although the program was begun on a shoestring at Houston Municipal Airport, it quickly developed to the point where it occupied its own facility at Sweetwater, Texas. To this day, Avenger Field stands as the world's only all-female airbase.

Notwithstanding different aims and purposes, the coexistence of the WAFS and the WFTD was decidedly uneasy. Within ten months of their founding, the two units were reorganized in a move that was a huge victory for Cochran. From the beginning, Cochran had argued that WFTD graduates should not be restricted to assignment with the WAFS. Indeed, by mid-1943 some were being assigned to different commands with very different duties. This weakened the key position of Love's group. It was therefore not surprising that when formal reorganization came, Love found herself in a diminished role. A new, single organization was created: the Women Airforce Service Pilots (WASP). At the top of this hierarchy was the Director of Women Pilots, a post to which Cochran was appointed. As such, she remained directly responsible for the training operation, but the designation WFTD was discontinued. WAFS, as a formal designation, similarly went by the board. Love continued in charge of the female pilots employed by the Ferrying Division of Air Transport Command, but her reporting line was no longer through that command; instead, it was directly to the Director of Women Pilots. All military airwomen were thus called WASPs from August 1943 forward.

The accomplishments of the WAFS, the WFTD, and the WASP are impressive. The training operation, for example, was an unmitigated success. The program attracted tremendous interest. Some twenty-five thousand inquiries were received from potential applicants. A total of 1,830 women were accepted, of whom 1,074 completed the program and accepted assignments as pilots. The "washout" rate, or proportion who failed, was similar to that of male flying cadets: data for 1943 show 26 percent for WASPs and 25 percent for men; corresponding figures for 1944, when all recruitment had dipped well below the top layers of the available pools, were 47 percent and 55 percent (Boom 1958, 531).

The military airwomen flew more than sixty million miles during the war, piloting seventy-seven different types of airplanes with safety records that exceed — often by substantial margins — their male counterparts' (Weatherford 1990, 68, 83). Their tasks were many. They towed targets for gunnery practice and simulated strafing and divebombing attacks as part of antiaircraft training. They flew searchlight tracking routines, towed gliders, test-flew planes being returned to service after repair, and checked out planes with reported malfunctions.

On several occasions, WASPs were used to demonstrate to male pilots the ease with which new types of aircraft or those that had developed a reputation for poor flying characteristics could be handled. The Bell P-39 Aircobra had a number of unusual characteristics because its engine was

Women Airforce Service Pilots M. Beard, J. Zuchowski, and M. Betzler at Romulus, Michigan, c. 1943. U.S. Air Force Photo Collection (USAF Negative No. B-29688 AC). Courtesy National Air and Space Musuem.

located behind the pilot. This fighter had developed a very poor accident record and hence a reputation as a killer. A WASP group at Romulus, Michigan, began ferrying the P-39 in 1943 and shortly after the accident rate for this plane declined substantially. This is thought to be due to increases in the confidence of male pilots, stimulated by the view, "If a woman can do it, I can do it" (Scharr 1986, 480–485). Female pilots were prominently displayed flying the Martin B-26, a twin-engine bomber with a bad reputation due to high landing speeds induced by small wing area. Even the B-29, the four-engine bomber introduced near the end of the war, was demonstrated by women in order to make its many technological innovations seem less formidable.

This essentially sexist use of airwomen was designed to shame men into better performance. It certainly continued a theme that had been at least an undertone in American aviation for decades. Some observers argue that the demonstration flights "boosted morale" of the male pilots in some way (Keil 1979, 256–258), but others believe they may have engendered resentment instead (Douglas 1991, 55). This matter will be discussed further in Chapter 5.

The most spectacular scheme for such morale boosting was General Tunner's plan to have Nancy Love and Betty Gillies, one of the original WAFS, fly a B-17 to England. The B-17 was a large four-engine bomber called the "Flying Fortress." There was great interest in this plane because it was the major weapon in air attacks on Germany; it was believed by many to hold the key to eventual victory. Huge numbers were ferried from the United States to England, particularly after mid-1943. The two airwomen were fully qualified for such a feat and in fact flew a B-17 to Goose Bay, Labrador on September 5, 1943, en route to England. General Arnold, however, did not share Tunner's view of morale boosting and so he stopped the transatlantic portion of the flight literally at the last moment. Regardless of the impact on morale, this flight would have been far more spectacular than Cochran's two years earlier. The urgency of the American military position, the size and complexity of the plane, and the fact that both cockpit seats were to be occupied by women would have made this flight newsworthy indeed. However, no champion with the clout of Floyd Odlum appeared.

Despite the outcome of this particular case, it is probably in aircraft ferrying that the WASPs made their most visible contribution. Indeed, ferrying was the original operational purpose of the WAFS, deriving from a strong experiential base. There were, however, many impediments to success even in this realm. It seems clear that General Tunner's consistent purpose was to employ the WASPs as effectively as possible in the discharge of his mission of moving planes. In order to accomplish this, he wished to advance his pilots, regardless of gender, to more sophisticated aircraft types as they gained experience. Many in the military, however, thought women should be restricted to less complex machines and above all should be protected from any possible exposure to enemy fire. The effect of such a restriction would be to leave Tunner with too many pilots for single-engine trainers and too few for multi-engine bombers bound for the battlefield. Accordingly, he fought the notion vigorously but with only partial success. Arnold's cancellation of the B-17 trip by Love and Gillies symbolized Tunner's consistent inability to move women to the very top of Ferrying Division's list by way of overseas assignments. This officer was equal to the challenge however, and with Love's help he began to move WASPs into a special role in the domestic ferrying of fighter aircraft. Most fighters were single-engine planes, but they were fast, powerful, and very heavily armed. These sophisticated warplanes required extremely competent, experienced pilots, but they were not generally ferried directly to operational bases in Europe or the Pacific. Because of their relatively short range, they often travelled to these locations on ships, but it was still necessary to move them from factories to various ports of debarkation. The WASPs, highly skilled but effectively banned from "war zones," fit the need for pilots perfectly.

Love had, by late 1943, established squadrons at four points in the United States. The airwomen found themselves under less intense scrutiny than they would have received at a single eastern site; they were thus better able to gain experience in fighter aircraft. Once the experience was gained, it was quickly put to use. Eventually 117 WASPs were employed in ferrying this class of high-performance aircraft. By fall of 1944, half of Ferrying Division's fighter pilots were women. Three-quarters of all domestic deliveries of fighter aircraft were being made by women. From the spring of 1944 forward, all Republic P-47 Thunderbolts, manufactured at Farmingdale, Long Island, and Evansville, Indiana, were flown by women on the first leg of their journey to war (Keil 1979, 243).

The collective WASP record is one of accomplishment and pride; the stories of individuals likewise reveal excellent performance and a willingness to endure hardship. Ann Carl was a recognized test pilot and was among the few to fly the P-59, America's first jet fighter. Barbara Erickson was awarded the Air Force Medal and a Presidential Citation for a ferrying mission that involved four flights, each over two thousand miles, in three different types of aircraft, over a period of five days. Thirty-eight WASPs sacrificed their lives in the line of duty. In retrospect the WASPs are viewed as courageous and patriotic, a group who went to great lengths to serve. In 1944, however, this was not the prevailing view. The WASPs were highly controversial; in fact, the organization was disbanded well before the end of the war. While the demise of the WASPs will be analyzed in a later chapter, it is instructive to consider the basics here.

By 1944 it had become clear that the war was winding down and that the United States would emerge as a victorious power. Though the fighting was far from over, the decision was taken to discontinue new starts in combat pilot training. The process of training a new combat pilot was lengthy, because of both the complexity of military aircraft and the long practice necessary to become proficient in their use. Individuals who would begin training in mid-1944 would not be ready to go into combat until a point at which, most experts agreed, hostilities would long since have ceased. The termination of new training starts had immediate implications. Pilot training contracts with private vendors were canceled. Many instructors suddenly found themselves without a job. These instructors were overwhelmingly male, generally too old for combat service, but still very far from retirement. In addition, many combat pilots were being returned to the United States, having completed the prescribed number of missions. The progress of the war ensured that they would not be called upon for an additional tour of duty.

Suddenly, the shortage of flying manpower that had justified the establishment of the WASPs no longer existed. Female pilots were no longer viewed as patriotic citizens who stepped into the breach so that men could be released for combat at the time of greatest need; rather, they were seen as usurpers who held jobs rightly belonging to breadwinners. Men needed jobs to support their families; women, presumably, did not. Women were represented as emergency substitutes, not as good as the genuine articles who were now once again available — despite the fact that the women were experts in flying the most sophisticated aircraft in the world, while most male ex-instructors had never handled anything more complex than a primary trainer.

The economics of the employment marketplace were translated into demands for government action. The Women Airforce Service Pilots organization was, with very little ceremony, disbanded. The WASPs were simply sent packing; since they were not formally a part of the military, it was not necessary to go through the procedure of a discharge. Indeed, from the beginning there had been disagreement about how to integrate this unit into the armed forces, and disagreement bred delay. The process was not completed by 1944, so the dismissed WASPs remained civilians, unable to collect even the price of a ticket home — to say nothing of veterans' benefits such as those male counterparts enjoyed.

The disbanding was justified mostly on the basis of cost, but the overt hostility toward women is evident in language used by a key Congressman: "To recruit teen-aged schoolgirls, stenographers, clerks, beauticians, housewives, and factory workers to pilot the military planes of this Government is as startling as it is invalid" ("Women: Unnecessary and Undesirable," 1944). A more thorough and complete rejection is hard to imagine.

Of course, emotion as strong as the Congressman's did not yield to any sort of factual consideration of the contribution that the WASPs actually were making. It is worth noting, however, that following the disbanding of the WASPs, Ferrying Division had to borrow 181 pilots from other commands in order to continue deliveries of fighter planes (Scharr 1988, 713). Although the WASPs had been sent home, the war was not over and someone had to do their jobs. Despite their impressive political victory and undoubted abilities with sixty-five-horsepower Piper Cubs, newly available male ex-instructors just could not manage P-47s.

Conclusion

On the face of it, the airwomen were an impressive group. From the very beginning, airwomen excelled as performers. Dramatic showmanship, aerial derring-do, and record setting were women's forte as well as men's. Airwomen's performance record is important in at least two respects. First, performing airwomen from Harriet Quimby through Mabel Cody to Amelia Earhart were actual or potential popular heroes. As such, they could have been, quite apart from the technical meaning of anything they did, powerful exemplars for many women. The airwomen had rare skills. They functioned in a new arena open only to a few. They faced danger and endured pain. Between 1912 and 1944, gender roles were undergoing some reexamination throughout the whole culture, a condition under which, perhaps, new and exciting role models could have some real impact.

Second, airwomen's performance feats were indeed, given the relative innocence of the times, astonishing and exotic. All members of the public, men and women alike, could hardly escape the fact that women were doing very unusual, very thrilling things. Moreover, performance caused the adrenaline to flow and this is exactly what the public wanted. There was a clear demand for the aviation spectacular; women obliged. Could beliefs about women's inferiority withstand this onslaught of factual evidence — for example, Thaden's victory in the Bendix? And even more important, could such beliefs withstand surges of emotion such as were caused by Smith's passage under the great bridges of New York?

The practical accomplishments of the airwomen were very real if somewhat less dramatically presented. The general public probably did not understand these very well, but aviation elites would have grasped them immediately. In an era of rapid and exciting growth of a new technological field, the practitioners might be expected to develop in-group loyalties capable of transcending traditional societal values. Something like shared professional norms, one would think, would have assured women a place in the increasingly influential aviation community.

Finally, the last of the great airwomen, the WASPs, introduced another element of accomplishment. Military service must be regarded as a type of practical contribution, but it also introduces the element of patriotism. On that account it is easy for both an aviation elite and the general public to appreciate and understand. The WASPs took on a fully military life complete with marching, living in barracks, low pay, and separation from loved ones. They generated performance statistics that met or exceeded any military criterion. They died for their country. Could an officer corps not accept such a decorated comrade as Barbara Erickson? Could a grateful citizenry do anything but weep for Evelyn Sharp, dead in the crash of a P-38 with a failed engine?

Dashing performance, brilliant practical contributions, and the flag seem like an unbeatable combination. While many people indeed admired the obviously admirable, we have already seen that the picture is more complex — and more discouraging — than one of great accomplishment followed by appropriate recognition. Airwomen had to rely more heavily on performance for recognition than their male counterparts, who had a wider range of professional opportunities. Were they, despite many practical contributions to aviation, somehow unable to shed the traditional image of the feminine performer? In the subconscious of the public, were they still of the circus or burlesque? In addition, there was a constant, ominous theme throughout this period: if women can pilot airplanes, then anyone can. This notion was used to change the image of aviation generally or to shame men into greater performance. There is an implication of feminine inability underlying this notion that, despite the record, may have continued to define the cultural norm.

It is interesting that as the 1912–1944 period progressed, airwomen's successes were based less on performance and more on practical accomplishment. That is, reality increasingly contradicted circus images and inferiority myths. But, it seems, resistance and opposition to the airwomen grew across the same time span. Helen Richey was hounded from her job with Central Airlines; by 1944 the WASPs were thoroughly despised in some quarters. Perhaps as the airwomen became more accomplished, they became more threatening. In any event, it is certain that the reality of the airwomen's accomplishments takes on particular meaning according to how they were perceived by the public and, more important, by influential segments of the society. It is to the comparison of reality and perception that we now turn.

❧ 2 ❧

Comments on the Conquest:
Promise Proclaimed

Although most Americans today are, at best, dimly aware of female aviators in the period from 1912 to 1944, the record is clear: women did indeed conquer the air. We already know, however, that they did not simultaneously conquer the hearts and minds of the nation — at least not to the extent that they forged permanent new roles for women in our culture. Somehow potential processes of education and inspiration failed to develop. Clearly, there were important lessons to be learned from the airwomen, but whether due to difficulties in their articulation, transmission, reception, or some combination of these, they were not learned. As a first step toward understanding what happened, this chapter will examine how the airwomen were received in their own time. Were their accomplishments clear from the outset or is their record something we can appreciate only with the benefit of nearly a century's hindsight?

Although the task is straightforward, it is very difficult to determine whether immediate contemporaries were impressed by the airwomen. We do not have good descriptions of people's specific attitudes during this era, let alone an understanding of what may have influenced them. Public opinion research was at a fairly primitive level then and few data are available that might help us. We should recall that one of the most sophisticated public opinion surveys of 1936 predicted that Alf Landon would capture the Presidency, when in fact Roosevelt scored an electoral landslide of historic proportions.

We can, however, look at the flow of information in the mass media. During this period, newspapers and other printed forms of communication largely carried the national agenda. If people were to be influenced — that is to say, educated or inspired — by a series of events, the publication process was central. We cannot say for sure whether anyone was listening, but we can determine what kind of messages were being sent.

Two features of the information flow are obviously crucial in any discussion of influence: volume and valence. Airwomen's lack of lasting impact on other women or American culture at large takes on quite different meanings depending on these factors. If the information flow had been minimal, lack of impact can be explained simply by the fact that messages were never sent. If there had been information flow but it was negatively canted, the best explanation would cite the persuasive power of the media. If a positive information flow had existed, the task of explaining the absence of influence becomes much more subtle.

In point of fact, there was a virtual tide of information about female pilots between 1912 and 1944, and the overwhelming majority of it was supportive. Though the presentation of airwomen had many interesting contours and a number of ironic twists, the overall image of them was one of dramatic, accomplished people who deserved respect. Their failure to influence society thus appears the more tragic, for they had a very promising beginnings.

This chapter examines four aspects of the communication flow: the writings of early female pilots themselves; the newspapers, magazines, and books of the period; movies of the day; and children's literature.

The Airwomen Speak

It is unlikely that the airwomen could have influenced the culture if they did not themselves speak out. In fact, many of the female pilots of this era were serious authors. Harriet Quimby, for example, was an accomplished professional journalist even before she took up flying.

To be sure, some airwomen wrote only after their era had ended. These works, typically autobiographies, are reflective and contemplative treatises rather than calls to action (Cochran 1954; Nichols 1957; E.

Smith 1981; Whyte and Cooper 1991). While the motives for writing them appear to vary from self-congratulation to an expression of gratitude for a good life, they are in no sense an attempt to exert leadership. Some airwomen however, also wrote for the same reasons as they flew. Though it is difficult to identify any of the prominent airwomen as the radical feminists or even committed activists of the day, it is clear that many understood the possible impact of their activities on women in the general public. These airwomen were fully conscious that their activities were unusual and dramatic for persons of their gender. They knew that if others lived by their examples, the culture would never be the same. In varying degrees they accepted this and wrote as advocates for a changed order. Leadership was surely requisite for the airwomen to impact the culture and many did not shrink from this responsibility.

There appear to be two main types of writing, each with a distinct message. Both are straightforward in their intent, but they nonetheless reveal some sophisticated insights. First, many of the female pilots thought aviation provided opportunities for women's professional and personal development. Many of these messages were addressed to other women and were clearly designed to be inspirational. Others sought to carry this theme to wider audiences. Though they were not necessarily given to bitter diatribes, the airwomen understood that society was shot through with stereotypes and prejudices that were not only restrictive of women, but had no basis in fact. They further grasped that men were the beneficiaries if not the architects of such beliefs; they understood well that women faced not only the general inertia of tradition but the active defense of male privilege as well.

Harriet Quimby characterized the macho personality in a way that might have described contemporary America. She thought that male pilots fostered "the impression that aeroplaning is very perilous work, something that an ordinary mortal could not dream of attempting" (Quimby 1912). One of her goals was to dispel that myth. Her opinion, and it certainly qualified as expert, was that flying took little muscular strength and was easier than most outdoor activities. Hence, she felt, it should be available to all, including women, both recreationally and professionally:

Flying is a fine, dignified sport for women, healthful and stimulating to the mind . . . there is no reason why the aeroplane should not open up a fruitful occupation for women. I see no reason why they cannot realize handsome incomes by carrying passengers between adjacent towns, why they cannot derive incomes from parcel delivery, from taking photographs from above, or from conducting schools for flying (Quimby 1912).

This was quite a message for women in 1912. This popular and successful role model was certainly not urging her readers to take gratification from the traditional arts of *Good Housekeeping,* the title of the magazine in which Quimby's article appeared. Articles of this genre continued to appear well into the forties.

The Stinson sisters were likewise conscious of general male domination. They saw their own activities as opening doors for women and believed in the necessity of direct competition with men. Katherine Stinson was very proud of her aerobatic innovation — executing a snap roll at the top of a loop — because no man had done it previously. In 1916 she impressed the aviation community by performing loops at night with flares attached to her plane. Clearly one to savor success, she was delighted that she had once again done men one better; she had, in her own words, "given them a bitter pill to swallow" ("Miss Katherine Stinson's Looping at Night," 1916). Marjorie Stinson was of a similar mind and consciously addressed other women. The motivation for her flying career was "to show some of the other girls how easily they can learn . . . flying. Flying, you know, never was intended exclusively for mere men anyway" (Stinson 1928).

Ruth Law was less directly combative in style, but nonetheless vigorously insisted on judging women on the basis of demonstrated competence rather than "theories" about innate, gender-related characteristics. She was widely quoted in this vein (see the discussion later in this chapter) but also took the time to prepare thoughtful articles. Law understood that airwomen were rekindling a "world-old controversy" by their success in the sky. She welcomed the controversy and planned to win (Law 1918).

There were elaborations on this theme. Some believed that aviation was unique, that it represented a historic juncture in gender relationships.

Marjorie Brown, a recreational pilot who was not known for spectacular record-setting flights, may nonetheless take the trophy for advocating aviation as the key to women's liberation. Her articles, while valuing new professional opportunities for women, more pointedly speak of abstractions and ideals. Women, she believed, "are searching for spiritual freedom"; and as she saw it, "flying is a symbol of freedom from limitation" (M. Brown 1929, 1930b).

The anticipated real-world results were left a little vague. Brown's enthusiasm may seem a bit naive, but she surely hit the issue of male-female relationships head on. The women who actually fly, she thought, will learn self-reliance, self-control, decisiveness, and other desirable characteristics to which they are now inadequately socialized. These characteristics will then enable women to achieve success more generally, to lead more fulfilled and competent lives, and thus to lessen dependence on men:

> Since time immemorial, man has been woman's protector. From the man's standpoint, the greatest satisfaction to be found in his relations with a woman lies in his ability to give of his protection, his largess, his care and her willingness to receive. . . . And it is an insult to their protective instinct for a woman to flout this proffered chivalry, to say by her actions, "Let me alone! I don't need any help. I can look after myself." This is just what a woman does say . . . when she gets into an airplane, opens the throttle, tosses her head, and goes sailing off alone into the sky. . . . Aviation points the way (M. Brown 1930a, 108).

These dynamics, plus the belief that nonflying women would also be inspired by female aviators, made airplanes, in Brown's mind, the great gender equalizer.

Only one full-length, comprehensive autobiography of an airwoman appeared as early as the thirties, that of Louise Thaden (1938). Although Thaden did indeed advocate professional and personal fulfillment for all women, this work is notable for its description of personal struggle. Thaden's aviation career was, like that of many of her contemporaries, an off-again–on-again affair. As we saw in Chapter 1, successes as a pilot were interspersed with periods of domesticity. Her restlessness caused all of her attempts at full-time family living to fail, except the last one, in 1938. In this year, she abandoned the aviation profession forever. She did, however, immediately begin work on her memoir, *High, Wide, and Frightened*. This book shows Thaden to be a very complex person. It is clear that even in retirement she never fully reconciled her urge to fly with the demands of a traditional role. *High, Wide, and Frightened* is much more than a chronicle of racing pilot adventures. It is a tale of frustrations and successes, of opportunities confounded by society's expectations. In it, Thaden outlines a whole series of prejudices against female pilots. The necessity of deferring to men with piloting skills inferior to her own left an especially painful mark. She ultimately came away with a very low opinion of the male ego and a high one of women's piloting ability.

Thrust incongruously into this autobiography is a fictional short story about female combat pilots in an imagined future war. The descriptions of her characters' joy and anguish expressed Thaden's view of gender roles as eloquently as her descriptive prose. But the most moving comments have to do with Thaden's frustration at domesticity and pregnancy. She greatly missed flying during her periods of commitment to family life, and she felt both oppressed and guilty. She suffered greatly in the role of following her husband, to whom there was never any doubt that she would take a back seat.

To be sure, *High, Wide, and Frightened* is not a story of revolt, much less the inspiring description of a successful one. But neither is it a chronicle of defeat and despair. It is an honest statement by the winner of the most prestigious long-distance air race in America that it is normal and even desirable for women to be unhappy with the strictures of conventional female roles. Thaden achieved some degree of self-actualization, and it was well worth the personal cost.

Some of the airwomen authored a totally different type of material. Since they were experts or at least knowledgeable participants, commercial publishers called upon them to comment on the whole range of aviation activities. Often, such writing addressed female aviators or gender roles minimally or not at all. The pilot adventure-travelogue was a popular genre. In *Aerial Vagabond*, for example, Bessie Owen describes what it was like to fly an open-cockpit plane from France to China (Owen 1941).

Among the most prominent authors of this kind is Ruth Nichols. Many of her articles deal with the overall piloting experience and simply describe to the reader what it is like to fly. Titles like "The Sportsman Flies

His Plane," "Aviation for You and Me," and "You Must Fly" reflect a generalized aeronautical boosterism (1930, 1929, 1933). In "Behind the Ballyhoo" (1932), Nichols sought to explain the motivation behind and meaning of racing and record setting. Many Americans were becoming somewhat dubious about the value of aeronautical adventurism and Nichols sought to address these concerns. This was not, in most respects, a gender-related issue. Nichols sought to describe and justify the personality of aviators; for the most part, these articles might have been written by a man as easily as by a woman.

The specific content of such articles is of little importance to our concerns, but the fact that such articles were authored by women is of the greatest significance. Women appeared to the public as experts on aviation. Women were qualified to speak about a new high-tech field. In magazines aimed at women and in those directed toward the general public, female authors spoke of a world far removed from home and family. This kind of writing demonstrated that aviation was not the exclusive province of men.

Also heavily involved in writing of this kind was Amelia Earhart. When it comes to communications, this airwoman constitutes a special case indeed. Earhart was in large measure a managed entity. Her writings were, in part, reflections of the established New York publishing elite. For this reason, she is treated in more detail in the discussion of the general press. Nonetheless, Earhart was a media figure of giant proportions and it is impossible to examine the potential impact of airwomen's writings without considering her work. Far more people read Earhart than any other airwoman.

As we have seen, Earhart rose to fame in 1928 following an aerial crossing of the Atlantic. *Cosmopolitan* immediately engaged her to write a series of articles and she became, in effect, that publication's aviation correspondent. It is notable that a major women's magazine would have such a correspondent and that the position would be filled by a woman.

Earhart at first wrote virtually monthly and continued to produce regular articles through 1932. Most of her titles bear marked similarities to those of Nichols. "The Man Who Tells the Flier, 'Go'" described early aviation weather forecasting (1929a). "Flying Is Fun" was straight boosterism (1932a), while "Your Next Garage May House an Autogiro"

Amelia Earhart in the cockpit of a high wing monoplane, date unknown.
Smithsonian Institution Photo No. 85-13246.

discussed the possibilities of mass ownership of helicopter-like rotary wing aircraft (1931b). Even "Mother Reads as We Fly" fails to discuss women in aviation and certainly does not take any kind of advocacy role (1931a); it merely observes that flying is so safe and serene that passengers may as well read to while away the time.

Twice, however, Earhart stepped out of this mode, writing fairly spirited pieces that urged women to excel and warned that masculine resistance could be expected. Following her solo flight across the Atlantic in 1932, Earhart was repeatedly told that she must possess exceptional courage for a woman. This stimulated an angry retort. In "Women and Courage," she downplayed her own personal willingness to take risks, but blasted the widely held assumption that women are, in general, inherently less courageous than men. Such a view she attributes to false masculine assumptions and to the fact that "through the centuries, women have been trained for other things" (1932b). In an earlier, more general article, Earhart laid out a fairly sophisticated socialization argument, noting that women were being squeezed out of aviation by the continuation of ancient prejudices (1929b). The piece is a ringing call to women to conquer their traditional fears and to demand full participation in the aerial age. Although these articles were far more vigorous than most of her works, Earhart clearly had qualms about publishing them, commenting, "Doubtless by now I am running the risk of becoming a heavy-handed feminist." One wonders whether Earhart on occasion escaped from her handlers and wrote — somewhat guiltily — what she really felt.

The airwomen were not only adventurers, but socially conscious, thoughtful observers as well. Many wanted their work to serve as a model for other women; they were anxious to hasten the process of cultural change. With considerable candor, some attacked the inequities of a male-dominated culture. Even the less aggressive staked out portions of the aviation enterprise for their gender.

Their thoughts committed to print are very different from — and more powerful than — mere positive commentary. This was advocacy and leadership. It is very difficult to blame the ultimate lack of influence of the female aviators on a failure in communication. With drama and feeling, they indicated a path. It must have been with great dismay that they saw it quickly grow over untrodden.

The Airwomen and the Press

While the airwomen themselves were impressive authors, their works obviously constituted only a minute fraction of the mass-media information flow. Indeed, any significance it might have had was dependent upon the context provided by the overall output of the publishing industry. In terms of sheer quantity of copy generated, the relationship with the media was a great success. Airwomen captured the imaginations of reporters, editors, and publishers to a truly stunning degree. Current newspaper stories, thoughtful articles in periodicals, and biographies all appeared in profusion.

The relationships that existed between female pilots on the one hand, and authors, reporters, editors, and publishers on the other, spanned an immense variety. The most striking was that between Amelia Earhart and George Palmer Putnam. As we have already seen, Putnam's interest in aviation was clearly motivated by the search for publishable manuscripts. Further, there is little doubt that Putnam, who was connected to the newspaper industry as well, fully understood the public appeal of women in aviation. When Earhart was plucked from obscurity in 1928, profitable book titles were on his mind. A woman was needed to generate publicity for the first female crossing of the Atlantic by air. The woman's abilities and skills were not an issue, since the actual piloting was to be done by two men who had already been signed on for the task. Earhart was selected for the role as passenger, duly earned the distinction of being the first woman to cross the Atlantic by air, and just as duly wrote a book about the experience, published by G. P. Putnam's Sons (Earhart 1929).

There has been much speculation about to the degree that Putnam, in his management and marriage of Earhart, was motivated by the thought of commercial gain for his enterprises. Though we will return to the Earhart-Putnam connection later, the point to be noted here is that Putnam presided over a vast outflow of published material regarding Earhart. In addition to newspaper and periodical articles, this included a general book about women in aviation (Earhart 1932), a book based on logs of the round-the-world flight (Earhart 1937), and a biography of Earhart by Putnam himself (Putnam 1939). The latter two items appeared after Earhart had been lost over the Pacific.

There is a very real sense in which Earhart was a creation of the publishing industry. She was certainly an unknown figure prior to 1928. Though she possessed a pilot's license, her efforts had not earned her any recognition and her professional involvement was not in aviation but in social work. Though Earhart was clearly a person of great courage and determination, there is considerable doubt as to whether she was a truly superior pilot. As noted earlier, her flying career is characterized by a number of accidents and near-misses — publicly explained away by the Putnam machine as the result of mechanical malfunctions. Many observers felt that she lacked fairly fundamental piloting skills (E. Smith 1981; Tate 1984) and even her original flight instructor remarked on her carelessness and her propensity to take chances (Southern 1974, 124). Her basic judgment has also been questioned, particularly with regard to some of the decisions made in connection with the round-the-world flight.

How can such an interpretation be resolved with Earhart's many successes? An answer may lie in technical superiority. Airplanes were expensive in the twenties and thirties, and many female pilots had great difficulty in accessing the newest and the best. Putnam's wealth might have partially offset less than first-class competitive abilities. For example, in the 1929 National Women's Air Derby, a transcontinental air race, Earhart flew a Lockheed Vega, a powerful, cantilever-wing cabin monoplane of the latest design. She took third in this competition, seemingly quite a credible performance, but she was handily beaten by Louise Thaden and Gladys O'Donnell, both of whom flew open-cockpit biplanes.

Considerations of ability aside, Earhart was an immensely successful media figure, and here lies her significance. Putnam's promotion of her created an ideal object for hero-worship. Everyone admired Earhart, except for a few technically knowledgeable members of the aviation community. These latter were not the people we would expect to take inspiration from Earhart or to emulate her.

Earhart had an enormous share of the press recognition accorded female pilots, but not even George Palmer Putnam could crowd other airwomen from the newspapers. There is no doubt that women and airplanes were a mix that sold both newspapers and books. Table 2.1 shows, by year, the number of items in the *New York Times* devoted to female aviators over the period of this study. The entries include news stories, features, and editorials. Earhart references are categorized separately.

Most of the data in this table fit very well with what we might surmise from other information. In 1927, the nation's interest in aviation simply exploded. This was a direct reflection of Charles Lindbergh's nonstop flight from New York to Paris, clearly one of the most dramatic aviation events, from a public attention point of view, in this century (Corn 1983, 22). The surge in press attention to women was in large measure a parallel to general attention to aviation.

Table 2.1
Number of Articles Dealing with Female Aviators
Appearing in the *New York Times*, by Year

	Earhart	Others	Total
1912	0	29	29
1913	0	11	11
1914	0	4	4
1915	0	0	0
1916	0	15	15
1917	0	13	13
1918	0	9	9
1919	0	10	10
1920	0	4	4
1921	0	14	14
1922	0	8	8
1923	0	2	2
1924	0	5	5
1925	0	3	3
1926	0	4	4
1927	0	75	75
1928	76	66	142
1929	16	100	116
1930	13	136	149
1931	17	153	170
1932	79	115	194
1933	32	82	114
1934	10	96	106
1935	60	95	155
1936	26	93	119
1937	123	57	180
1938	24	38	62
1939	7	38	45
1940	1	34	35
1941	0	19	19
1942	2	61	63
1943	1	24	25
1944	2	31	35
1945	1	7	8

*Raw data used in the preparation of this table are found in Niekamp 1980.

The variation in the number of Earhart entries is in accordance with what one would guess from common knowledge. Putnam surely saw the 1927 upsurge in media attention to female fliers; 1928 was the ideal year to fly a woman across the Atlantic and reap the press benefits. Escaping from the role of passenger, Earhart indeed made a solo crossing of the Atlantic in 1932 and flew from Hawaii to California in 1935. Both flights were firsts for a woman. Press response generated the larger entries for those years. Of course, Earhart's 1937 around-the-world flight riveted both public and press attention on Earhart. Though Putnam had meticulously planned for the mass media to record triumph, the media machinery, once in place, was equally efficient in chronicling tragedy and mystery.

More generally, it is clear that from 1927 through 1944, women's exploits in the air were indeed covered by the press. Throughout this period, there was an average of two articles per week about female fliers and, in some years, the average approached four per week. It would be very hard to argue that the media manipulated coverage to prevent attention to feminine achievements in this field.

Of course, while this absolute level of attention seems impressively large, the question can be asked, "Compared to what?" An examination across time partially answers this question. Women's participation in aviation declined dramatically after 1944. One would expect little press attention and few articles in the *New York Times* simply because, compared to the twenties and thirties, relatively little was happening. This was precisely the case. Taking the article count at five-year intervals from 1950 through 1975 reveals numbers of fourteen, sixteen, five, nine, four, and six. The 1927 through 1944 numbers thus are large compared to a period of low female activity in aviation. When there was something happening in women's aviation, it was reported. It is interesting that jazz, a musical form of great importance to the popular culture of the period, received far less attention than did female pilots in the *New York Times* during this period.

To be sure, systematic content analyses of other newspapers have not been carried out, but other studies of the period suggest that intense press coverage of women in aviation was not confined to New York (Corn 1983; Potter 1985).

Periodicals other than newspapers did not hesitate to jump on the same bandwagon. Articles covering female pilots and their activities numbered in the thousands. The most popular sorts of magazines generated much of this copy; *The American Magazine* averaged an article per year in the period from 1929 to 1944. Those journals appealing to smaller and more specialized clienteles, as diverse as *Country Life*, *Living Age*, and *The Nation*, also published several articles each during this period. It is clear that the American public was exposed to a very generous information flow regarding the female pilots of the day.

It should not automatically be assumed, however, that media coverage was a good thing for aviation generally or for female fliers in particular. As will be discussed later, media promotion was actively and successfully sought in ways that can only be regarded as exploitive. Media stereotypes may even have encouraged subservience of women in the aviation community. Wing walking, a feminine role that persists into the airshows of the 1990s, comes immediately to mind. The publisher's spotlight can turn out to be, at best, a mixed blessing.

Although the total picture is one of subtlety and texture, it is clear that the publishing industry not only attended women pilots, but it also presented them in an overwhelmingly positive light. To be sure, the years between 1912 and 1944 were ones of massive change. Aviation itself matured greatly, moving from a mechanical curiosity to a huge transportation industry. Further, society itself underwent dramatic upheavals, including the extreme economic dislocation of the Great Depression and the most extensive war the country has ever known. The nature of thinking and commentary about any meaningful topic would be expected to change accordingly across the time span. Indeed, female pilots were viewed quite differently in 1944, as compared to 1912. Yet even as society appeared to periodically reassess airwomen, even as the fortunes of female pilots waxed and waned, the press by and large continued to offer support.

The earliest years of aviation were surely those in which the press was most completely supportive of female aviators, and for that matter, of aviation personalities generally. About the true pioneers, nary a harsh word was said, and women's achievements seem to have been equally as well received as those of men. Even though the sinking of the *Titanic* took up whole sections of newspapers for many days, the *New York Times* found

Harriet Quimby in a Bleriot Type XI monoplane, c. 1912. Smithsonian Institution Photo No. A 44401 C.

space for both an editorial and an article (albeit on page 15) about Harriet Quimby's 1912 English Channel flight. While it may be difficult for a newspaper to wax enthusiastic in an issue dominated by reports of one of the greatest maritime tragedies of the century, Quimby is clearly regarded as a dashing and important figure. In a discussion of her Channel flight and her overall career, readers were told that "this particularly daring young woman has led the way for her sex" ("Miss Quimby Flies English Channel," 1912, 15). Quimby's death in an airplane crash later in 1912 was the occasion for praise in a number of publications, from newspapers to such august journals as *Scientific American* ("Fatal Accident in Boston," 1912).

Ruth Law's nonstop flight from Chicago to Hornell, New York, was very widely praised. *Scientific American* offered plaudits because of the breaking of a record and elevated her feat to even greater stature because of the obsolescent machine that Law flew ("Ruth Law and Her Remarkable Flight from Chicago to New York," 1916). *Literary Digest* called her the "New Crowned Queen of the Air" and wrote at great length about the details of her trip ("New Crowned Queen of the Air," 1916). On the day following Law's record, the *New York Times* published nine items on the "finest feat of the year." These included a center story on page one, a long telephone interview, and an editorial entitled, "A Woman Flies 590 Miles" (1916). This piece is effusive in its praise, asserting that the event "shows how completely Miss Law has mastered the difficulties of flying and how trustworthy the aeroplane has become."

While this early commentary clearly remarks on the gender of the pilots involved, there is an almost matter-of-fact acceptance of women as participants in the emergence of aviation. There is no suggestion of special roles for women and, indeed, there seems to be the full anticipation that women are to be part of the aeronautical future.

This accepting attitude is found even when moves from the general press to specialized aeronautical publishing. To be sure, there were very few aviation periodicals prior to 1920, but it appears that the nascent industry gave due attention to women. The weekly *Aero*, published at least between 1910 and 1912, routinely reported the exploits of female pilots without particular emphasis on gender. Several cover photos featured airwomen of the day. Ruth Law was clearly regarded by the early journal *Flying* (not connected to the current high-circulation magazine by that name) as one of the notable aviators of the time ("Ruth Law's Record Breaking Flight," 1916; "Miss Ruth Law Presented With $2,500 Purse," 1917). *Aerial Age* was greatly impressed by the accomplishments of the Stinson sisters and regularly featured them in its commentary (See photo, p. 85). This publication did note the small stature and relative youth of these airwomen, but it did not regard such variables as disadvantages that had to be overcome. *Aerial Age Weekly* spoke of accomplishment and not gender ("Katherine Stinson in San Diego," 1915; "Youngest Flier in America a San Antonio Girl," 1916). Perhaps aviation at that point in history was so new that gender-based expectations of the greater society had not yet penetrated it. Perhaps its practitioners were so few that any sort of exclusivism seemed pointless.

Katherine Stinson on the cover of *Aerial Age Weekly*, 1915. Smithsonian Institution Photo No. 76-1140.

While the naive enthusiasm of aviation's earliest years continued in some sense for a half century, two themes with major implications for women were laid over it. One, discussed in Chapter 1, was very general. Beginning about 1920, many began to think seriously about the long-term future of aviation. Though daredevil adventurism was to remain part of aviation for years to come, additional potentials, particularly commercial ones, took on new importance. There were calls for aviation to become responsible, reliable, respectable, and safe. Though these concerns did not exclusively address women, safety in particular was to have a special meaning for them.

The second theme, by contrast, specifically focused on women. Progressivism, suffrage, and Prohibition — major if not revolutionary forces in early twentieth-century America — all involved women in unprecedented ways. Women had successfully called attention to themselves and had begun to work out new roles. In some cases, such as voting, these roles stressed equality with men, but more often they spoke of feminine distinctiveness. Women needed protection in the workplace; women's superior morality was needed to curb men's inclination to drink. The search for unique niches spilled over from the general culture to aviation; by the late twenties, airwomen were sometimes represented as having different sets of motivations from men. The consequences were substantial. Both of these themes and the concerns that they raised remained current through 1944 and arguably well beyond. The publishing industry, to its credit, handled both without retreating from its support for female pilots.

Aviation matured greatly during the era of the airwomen; its commercial development proceeded by leaps and bounds. However, aviation never completely shed its heritage of performance and exhibition. The resulting dualism led to serious tensions in some cases, more or less successful resolutions in others. To many, barnstorming and then later aerobatics, racing, and record setting seemed bumptious and inappropriate to an industry that held great promise for transportation and commerce. Calls for transportation development activities were accompanied by demands for an end to setting records for endurance, speed, distance, altitude, or number of consecutive loops. Some felt that such activities was inherently dangerous and others argued that they had no substantive purpose other than self-serving publicity.

On the other hand, defenders insisted that such spectacular feats contributed to the knowledge-base of aviation in a rigorous if unsystematic way. Some observers felt that "stunts" and other heroics were immensely valuable because of the public attention and support they drew to aviation; in an era when the future was thought to depend on airmindedness, this was no inconsiderable claim.

This dichotomy agitated the aviation community, of course, but even the daily press reflected the split. Often there were hotly written articles in long series. Perhaps one of the most notable was a ten-article sequence on stunt flying that appeared in the *New York Times* from January through September of 1930. It must be admitted, however, that widespread public discussion did little to dampen the enthusiasm of aeronautical daredevils of either sex.

By 1928, when Elinor Smith flew under all of the East River bridges in a single flight, such feats drew more than excited praise. Smith's effort, though it required remarkable concentration was foolhardy. Air currents were unpredictable and the flight had to be carried out at such an altitude that collision with barge or ship traffic — to say nothing of towers, floors, and cables of the bridges themselves — was distinctly possible. Though Smith was not arrested or charged, she found herself, a few days after her audacious adventure, in the mayor's office. She endured a severe lecture on public safety, muted somewhat by that official's clear admiration (E. Smith 1981). The same duality showed in the press. The *New York Times* reported the event as a feat that had never been accomplished before, but a few days later delivered a critical editorial upbraiding her for irresponsible and unsafe flying ("Says She Flew Under East River Bridges," 1928; "Girl Pilot Is Punished: Her Sentence Is Mild," 1928).

While Smith probably would have been chastised regardless of her gender, it appears that female aviators were special objects of such criticism. A new emphasis on safety became an excuse for woman-bashing, presumably because the exposure of those delicate creatures to danger is far more distasteful than risks that rugged men might experience. Ruth Elder, a Hollywood actress who was also an accomplished flier, generated a massive outcry in 1927. Serving as copilot, she attempted to fly the

Ruth Elder, well-known actress, and the WACO Model 9 in which she learned to fly, 1927. Smithsonian Institution Photo No. A 51327.

Atlantic with George Haldeman. Had they been successful, Elder would have upstaged Earhart and Putnam by nearly a year, but the flight ended in a ditching near the Azores. Elder and Haldeman, after a loss of oil pressure, had to put their big monoplane down near a passing freighter. The plane exploded and sank but the two lived to tell about it. This near-disastrous failure earned Elder a great deal of criticism for her presumed motives: promotion of her movie popularity and enrichment of her sponsors.

To be sure, Elder was probably not unaware that being the first woman to cross the Atlantic by air would have positive implications for her screen career. Indeed, she was able to realize a great deal of publicity anyway. The rescue at sea was dramatic and well reported; fortunately, some of the freighter's crew had cameras. Similarly, it would be naive to think that the flight's various backers were uninterested in possible commercial spin-offs.

Nonetheless, commentary on this event was in striking contrast to that discussing Lindbergh's flight of only a few months before. One observer professed to support women fliers, "but when they plan a feat, risking human lives, and then seek to make of it a money-getting scheme, the public grows suspicious." Scarcely less scathing is the assertion that flights such as Elder's should not be undertaken "without compelling reason and without the most thorough preparation" ("American Supergirl and Her Critics," 1927). While Elder's intent and level of responsibility certainly took a drubbing, she had many supporters, particularly among the more thoughtful commentators. The *Literary Digest* was mildly offended by her detractors. *The Nation* went on at some length to explain motivation in such cases and, through a realistic discussion of the issues involved, effectively defended Elder from her critics (Owen 1927).

Beginning in the twenties, many aviation publications paralleled the popular press in downplaying the spectacular and the bizarre. Indeed, their constituencies had the greatest stake in the development and maturity of flying. Their more reasoned and informed pursuit of responsible aviation, far from rejecting women's involvement, included them as essential participants in the process. Well into the thirties *Popular Aviation* featured at least two columns devoted to female pilots: "Aero Sportswoman," by Joan Thomas, and "Women's Activities," by Lady Mary Heath, a famous British airwoman who maintained many American connections throughout her career. Heath was a record setter herself, and her view that women as well as men should be testing their wits and their machines' capacities in full competition came through loud and clear. Only through such challenges, Heath believed, could aviation develop the sophisticated knowledge required for real advances. Although the appearance of specialized columns for women perhaps suggests the beginning of segregation to a peculiar status, the content of the columns and their fit within the general parameters of the magazine do not support this notion. Women seem to be treated as full citizens of the aeronautical community.

In the twenties and thirties *Aero Digest* published short biographies of many of the airborne personalities of the day. Female subjects included Ruth Elder, but also lesser-known figures such as Florence "Pancho" Barnes, a particularly colorful air racer who was also featured in a made-for-television movie in 1988 ("Ruth Elder," 1928; "Florence Lowe Barnes," 1930). This periodical did not eschew aviation safety. On the contrary, its thrust was to defend the right of women to take the risks inherent in aviation along with men — and thereby to contribute to a safer and more reliable transportation industry in the future.

Of course, accidents involving women, particularly fatal ones, drew the greatest public outcry. While female pilots' exploits attracted special attention for a thrill-seeking public, the death of a woman engendered particular revulsion. Blanket condemnations of women's flying abilities and demands to banish them from the skies were not uncommon. These are quite different from the sorrow and mourning that attended Harriet Quimby's fatal dive into Boston harbor. Even in the case of fatalities, however, much of the press continued to support women as full-fledged participants in aviation.

Perhaps the most notable example of fatality-related criticism involved the 1929 National Women's Air Derby mentioned earlier. This was a Santa Monica–Cleveland race flown solo by twenty experienced female pilots over the course of several days. It was immensely well publicized, a media event of the first magnitude. Many of the participants were quite well known. They included actress Ruth Elder, dried out from her dunking in the Atlantic two years previously, and Amelia Earhart, who was by then an object of considerable promotion by the Putnam publishing enterprise, having already been flown to England in 1928.

An atmosphere of festive excitement was expected to characterize the week-long National Air Races, of which the Derby was one event. It was marred, however, by the death of Marvel Crossen, a commercial pilot from Alaska. The accident occurred only one day into the race and there was widespread demand that the remainder of the event be canceled (Brooks-Pazmany 1983, 34–51). Crossen's crash allegedly showed that women just were not competent to navigate the skies — despite the fact that she had spent years successfully negotiating the real rough-and-tumble of arctic bush flying. One observer, in complete disregard of the evidence, claimed that Crossen had neglected to open her parachute after bailing out (Corn 1983, 79). Ultimately, however, the press seemed more impressed by the fact that fifteen of the twenty entrants completed the race course; recognition for a great deal of determination and skill was duly given.

On the occasion of at least one fatal crash, the *New York Times* took special pains to blunt popular criticism and defend the role of women in aviation. Laura Bromwell was a successful commercial pilot and exhibition flier in the New York area. She supposedly once completed 199 consecutive inside loops. She was widely popular and participated in law enforcement activities of the aviation division of the New York Police Department. Her death in the crash of a new aerobatic plane thus raised some outcry, particularly as she may not have been properly seat-belted. The *Times* editorial, however, must be regarded as quite remarkable for 1921:

> There is little if any reason for assuming that Miss Bromwell as an aviator was less competent than a man of the same training, experience, ambition, or taste. . . . All fliers are so wonted to the taking of risks and so frequently do they go aloft when conditions are not perfect and the principle of "safety first" would keep them on the ground that Miss Bromwell's failure to have her holding strap in perfect order should not be taken as counting against women as a class to compete with men in this new profession ("Women as Flyers Considered," 1921).

Even Amelia Earhart, by 1937 a figure of great popularity and stature, was not immune from attack. There were a few unkind comments about her following her disappearance over the Pacific. *Time* magazine regarded her attempt to fly around the world as a useless "stunt" and stressed that all future such activities should be strictly prohibited ("Search Abandoned," 1937).

Although this article echoes other reactions to the deaths of other female aviators, its sentiment was unusual in the wake of Earhart's disappearance. Earhart was widely praised and there was a great outpouring of sympathetic copy. Most of it not only praised Earhart personally but also noted how her career — including the events surrounding the ultimate tragedy of her death — had greatly enhanced the future of aviation ("Amelia Earhart," 1937; "Clouds and Victory," 1937).

In short, during the twenties and thirties the meaning of aviation changed. No longer was uncritical acclaim given freely to pilots, male or female. Sheer adventurism was regarded with an increasingly jaundiced eye. The demand for safety did in some quarters focus heavily on women, those who traditionally have been thought to require protection from danger. Most of the press, however, remained not only positive, but took female pilots' detractors to task. Often women were shown to be key players in the development of the mature, reliable aviation industry that the critics themselves professed to desire. Female fliers were still regarded as heroes.

The second developing theme during this period, the identification of airwomen as motivationally different from men, was much more complex and potentially much more serious. Of course, female pilots had always in some sense been distinctive; discussion of distinctiveness in terms of the compatibility of flying with basic women's roles, however, was an innovation of this period.

This phenomenon was very subtle. Its development did not require overtly negative attitudes toward women. While in some cases there were strident demands that women stop flying and start gestating and lactating instead, more commonly there was simply an expression of concern with how prominent airwomen handled matters presumed to be of uniquely feminine concern — for example, fashion and family. While such concern may even have been consistent with support for any individual female aviator, it signified the vitality and persistence of a set of cultural values that have long restricted women's achievement of full citizenship.

The clothing and makeup worn by female pilots, as well as their general appearance, had always intrigued the public, but only from about 1927 is this interest prominent. Articles about fashion shows for female pilots presented very detailed descriptions of articles of clothing and the way they were modeled ("Women Fliers Act as Fashion Models," 1935). When Amelia Earhart undertook to capitalize on her fame by producing a line of sportswear, the press could not resist commenting. Earhart's skills as a designer and the details of the various items in the line made good copy indeed ("Designs Flying Clothes," 1933). What the well-dressed male aviator might choose for his wardrobe was never a matter for the newspapers.

The WASP received an extensive photo treatment in *Life* magazine on July 19, 1943. The article was by no means hostile; in fact, it acknowledges these airwomen's patriotism and dedication. However, as Luanne C. Lea argues,

> The *Life* article projects a female stereotype who is especially concerned with her appearance. [It] implies that the major problems in flight training are unattractive coveralls and varieties of headgear that will control but not disrupt hairdos as, "feminine locks constantly creep into girl fliers' eyes." *Life* closes with a caution that the calisthenics program "will only strengthen strategic muscles in fliers" . . . no unsightly disfigurement will occur (1992, 3).

The domestic relations of airwomen became an even more prominent concern than fashion. Consistent with the idea that women should be particularly involved with family, an image developed of airwomen who were in some sense tributary to or dependent upon a husband. Commentators implied that women may take up flying only when their husbands permit it, and that a female pilot's activities should be useful to her husband's career or business. The assumption was that whatever women in aviation do, it must support rather than threaten their spouses.

The classic case is Anne Morrow Lindbergh. Having married the ultimate aviation hero Charles Lindbergh in 1929, Anne Morrow became an accomplished pilot in her own right. Her greatest contributions, however, were as an author. She ultimately wrote a number of books, but those that appeared in the thirties and are the best known promoted the idea of airwomen as dutiful helpmates to their flier husbands. Both *North to the Orient* (1935) and *Listen! The Wind* (1938) emerged from extended flying trips she and her husband had undertaken to popularize aviation and encourage the development of long-distance air travel. Both were tales of adventure and exotica, a typical genre of the day. Both became best-sellers.

It is quite possible that these books contributed to the airmindedness of American citizens and they certainly deserve the literary praise that they received. However, they conveyed a dual message for women. First, in all things relating to the technical aspects of aviation, Anne Morrow Lindbergh clearly deferred to her famous husband. While she acknowledged the creative sensitivity necessary to chronicle their adventures, she disclaimed any degree of aviation expertise or skill.

Indeed, Anne Morrow Lindbergh's degree of self-effacement sometimes approached the ludicrous. She acted as radio operator on the couple's joint flights to the Far East and around the Atlantic. To be sure, early aviation radios were somewhat less user-friendly than basic communications gear of today, but the situation was hardly overpowering. To change frequencies, one had physically to replace a coil. Most transmissions were keyed in Morse code rather than spoken into a microphone. Anne Morrow Lindbergh made a great fuss about her initial ineptness in this role, the difficulty of learning such a male activity, and how indebted she was to her husband for helping her to learn the craft. She was apparently immensely grateful that he eventually assigned her the role of responding to reporters about radio communications: "My husband never answered any questions about radio even when he knew the answer far better than I did. He would just turn to me with the expression half proud and half anxious that a mother wears talking to her performing child, 'Speak up now, Anne, say your piece for the gentlemen' " (Lindbergh 1935, 51). While such words would have seemed quaint to airwomen who had mastered all aspects of flying, including radio communications, it carried a different implication for female readers who were ambitious and curious but uninitiated.

Some traditionalists during the twenties and thirties worried about female fliers' responsibilities to children and husbands as well as their more general obligations toward homemaking. Obviously, Anne Morrow Lindbergh neither symbolized nor argued that a woman's place was in the home. She was heavily involved with aviation and did not reflect a traditional homemaker's lifestyle at all. On the other hand, she was anything but a feminist — indeed, her involvement in aviation was a direct result of demands from her husband. Charles Lindbergh rejected tradition in that he wanted a co-participant aviation buddy for a wife, but he was very traditional in that he expected his wife to adjust to his demands in becoming just that (Mosley 1976, 145). While her overt activities seemed in substance spectacularly liberated, Anne Morrow Lindbergh was, in fact, a dutiful wife.

This is evident in her substantial transformation following marriage. Prior to marriage she was an active person on the edge of public life; the Morrow family was prominent in government, her father serving as ambassador to Mexico in the late twenties. She was intellectually inclined, a reflective writer of poetry. After marriage she became an aeronautical Sancho Panza and something of a recluse, adopting her husband's detestation of the press.

Anne Morrow Lindbergh's books clearly reflect her subservient posture. She is always in the rear cockpit receiving instructions, wondering what decisions Charles would make, and she was always helping out. Nonetheless, trips "north to the orient," that is, following the great circle route to the Far East, were so pioneering that her reports earned a number of awards, including a major medal from the National Geographic Society. These recognitions spoke dramatically of what might be expected of successful airwomen in America.

Anne Morrow Lindbergh was not alone in being identified with an aviator spouse, nor were books the only medium through which this connection was made evident. Early on, women were recognized in the press simply for accompanying their husbands on notable flights ("Mrs. A. Cobham Will Fly With Husband From England Around Africa," 1927; "Mrs. J. L. McDonnell to Go As Passenger in Race," 1927). Later, even nonflying husbands were brought into the picture to share in awards earned by women for aviation achievements. So accomplished a figure as

Jacqueline Cochran was sometimes not recognized on her own but as part of a married couple ("Gets Rumanian Air Award," 1939).

Of course, many of the early airwomen were married, and a significant proportion of their husbands were also pilots. In 1932, one of every five female pilots in the United States was married to an aviator (Corn 1983, 82). This attracted considerable attention and favorable commentary, as it seemed to reaffirm the ideals of marriage and family. Some writers clearly thought that husbands did the teaching and wives did the learning about aviation, that the Charles Lindbergh–Anne Morrow Lindbergh syndrome could be generalized to other couples ("Husband-Wife Teams in the Flying Game," 1930).

In the 1930s, Americans were intrigued by Amy Johnson, a British pilot well known for pioneering flights from England to Asia and Africa. To be sure, she did receive a great deal of attention for her accomplishments, particularly for a 1932 record-time solo flight in an open-cockpit biplane from London to Australia (C. Smith 1967). But the attention devoted to her was greatly enhanced by her engagement and marriage to, air adventuring with, and separation and divorce from a British flying playboy, James Mollison. Mollison was a flier of some note, having crossed the Atlantic several times. He and Johnson eventually completed a partially successful flight from England to the United States (it ended with a crash in Connecticut rather than a landing in New York) and flew together as a team in the famous 1934 MacRobertson race from England to Australia. While the press did not suggest that Johnson was less capable than or dependent upon Mollison in these joint ventures, it most assuredly did dilute the impact of her competence as a flier by intermixing it with news about her romantic involvement (or lack thereof) with her husband. Out of seven *New York Times* stories about Johnson in 1937, three described her deteriorating relationship with Mollison and four were about aviation.

While Johnson had the misfortune of being associated with a particularly visible man, one who made a reputation by participating in one international scandal after another, she was not the only female pilot whose domestic relationships attracted the press. Female pilots were associated with marriages, divorces, and even bigamy in a way unparalleled in discussions of their male counterparts ("Ruth Alexander Death Discloses Doubts of Marital Status," 1930; "Lady Heath Expected to Receive Divorce Decree," 1930). In fact, the latter's domestic circumstances were largely ignored.

Occasionally, the concern with men in the lives of female fliers reached faintly ridiculous proportions. In 1935, George Palmer Putnam had used the word "baby" in referring to difficulties in certain business situations. This statement was progressively misquoted and taken out of context until it was reported as the demand of a frustrated husband to a wife unwilling to bear him a child. Of course, the last thing that Putnam needed in his promotion of Amelia Earhart was the image of someone badgered by her husband — much less the actual birth of a child — but even his influence in the publishing industry could not stop such rumors. Earhart was obliged to spend some time in earnest conversation with reporters in order to explain the matter away ("Earhart Gets Advice on Babies," 1935).

Women's involvement in aviation was portrayed as distinctive or unusual in a host of other ways as well. Some stories were downright weird. Much was made of Beulah Unruh, a sometime waitress who ultimately earned a pilot's license. Allegedly, waitressing involved much standing and walking with the result of painful feet. She learned to fly in order to get her feet off the ground so that she could achieve some relief (Stockbridge 1931). "I Let my Daughter Fly," proclaimed one article; although it thoroughly approved of such a decision, the article differentiated daughters from sons and raised the question of whether young women needed special protection from the allure of flying (Beatty 1940). The validity of old maritime taboos against women and their extension to aviation were seriously discussed in articles like, "Do Women Hoodoo Transatlantic Flights?" (1931). The presence of a woman aboard a sailing ship was once considered bad luck by crew members. Some observers in the thirties, presumably more knowledgeable if not more intelligent than yesteryear's deckhands, purported to show that airplanes carrying women were less likely to complete an ocean crossing.

If women are considered distinctive in motivation or character, it is but a short step to suggest that they should occupy distinctive roles in the aviation community. Indeed, right on schedule, two such roles emerged, and the thrust of each was to exclude women from piloting altogether.

Those women of the late twenties, thirties, and forties who responded to these particular expectations and occupied distinctive roles may have had interesting lives, but they were surely much more passive ones than those spent in the cockpit.

The first role is acting as a passenger, simply being flown from place to place. Amelia Earhart's crossing of the Atlantic in 1928 is a perfect example. A number of other women achieved some degree of notoriety by riding in early airplanes over some unique or interesting route. Attention to such travels was intense but of short duration, and "passenging" as a notable activity seemed about to die out as aviation matured.

By the early thirties, however, fairly extensive domestic and foreign airline route structures had begun to develop. More systematic and regular commercial air travel had an appeal beyond the practical function of getting to a desired destination. Going by plane, even scheduled airliner, was still rare enough that travelogue books based on air journeys had a fairly extensive readership. One could thrill to accounts of a tour by flying boat around Latin America or feel vicarious wonder at other faraway places with strange sounding names that were now at least marginally accessible by plane. Many of these books were written by women (Cranston 1935).

While passenging was very different from piloting, it did require some initiative and a taste for adventure. Beginning in the mid-thirties, women managed five-continent trips by stringing together the routes of various scheduled carriers (Shuler 1939). The ultimate, perhaps, occurred when Clara Adams made commercial air travel a competitive event. Much after the fashion of Phileas Fogg, she set out to go around the world from Newark to the east. Hopping one airliner after the other, she completed the circuit, arriving back at Newark sixteen days and nineteen hours later. She claimed a world record, and though this was 1939, many observers were impressed. Regrettably, some commentators noted that women certainly make better air passengers than men, because they have more leisure ("Women in the Air," 1939).

In 1930, United Air Transport began a practice that carved out a second new niche for women in aviation. That carrier, quickly followed by Eastern, began to install "hostesses" or "stewardesses" on its passenger flights. Both magazines and newspapers were quick to comment on this development, and soon the role of women in aviation was substantially redefined.

The innovation of hostess service was yet another variant of an already mentioned brand of aeronautical sexism. The airline industry needed to blunt the still-present public anxiety about flying. If passenger travel was to become common and therefore profitable, the public could not continue to view flying as a special adventure for the hardy and daring few. In order to transform the airliner cabin into a comfortable, nonthreatening environment, the airlines supplied a nurturing female — a stewardess, who in the early days was also a registered nurse. The importance placed on female nurturance becomes evident when one considers that the airlines did not employ stewards, a logical parallel to maritime travel, or something akin to railroad porters, who were always male.

With the advent of stewardesses, interest in women in aviation became blatantly sexist. Although the previous reaction to women in aviation was not always gender-neutral, assertions that stewardesses must be "useful as well as decorative" and claims by airline executives that physically attractive women in airplane cabins would lure male passengers explicitly sound the theme of exploitation. Exploitation of women in aviation occurred in a number of contexts, but blatant admission of it — and even attempts to justify it — was unique to the stewardess role. Stewardesses were, like today's flight attendants, service personnel (Cleveland 1935; Shuler 1936). All in all, however, it seems that their jobs were demeaning. They held passengers' hands, laced up their shoes, and in general behaved "like a mother trying to keep the children amused" ("Hostesses of the Air," 1932).

Stewardessing represents the first concerted attempt to recruit women into an aviation career. To be sure, observers of women in aviation had for many years claimed that the future was bright for female pilots. However, the active, "this is just right for you" message directed at potential stewardesses was new: "There is no job on earth so attractive to the American girl as the one that takes her off it" (Drake 1933, 187). Discussions of stewardess qualifications, numbers of applicants, and the hiring process were common. Systematic direction on how to become a pilot was not easy to find; by contrast, guidance to the stewardess role was available without even asking. An aviation career for women had at last been found (Mackenzie 1937).

Commentators were quick to point out that stewardessing had a tremendous advantage over other careers for women: the opportunity to meet and ultimately marry a desirable man in the person of a pilot or passenger. Passenger airplanes "became the greatest matrimonial bureau known" (Hageb 1937, 1). Moreover, this new career was carefully distinguished from piloting and placed in clear subservience to the males who occupied the cockpit. As one stewardess put it, "There aren't going to be any girl air mail or passenger pilots. . . . By taking our home-making instincts into the . . . airliners, we can lend familiar aspects to which travelers may cling" (Courtney 1932, 29). Not only would women learn useful skills like neatness and meal preparation while serving as stewardesses, but they would also experience the gratification of an age-old relationship: "Win or lose, the [stewardess] sticks by the man up front. His dangers are her dangers. Her life is in his hands, and that's all right with her too" (Drake 1933, 193). This image of stewardesses, which was to persist for at least fifty years, was very different from that of airwomen who flew in the 1929 National Women's Air Derby.

Once a clear aviation role had been defined for women, the hostility toward those who sought to occupy other roles increased. While overt disapproval of female pilots was rare in the earliest years, there are some notable examples after 1928. One particularly fierce diatribe asserted that women lacked the discipline and concentration necessary to operate a complex machine like an airplane; when this was coupled with their forgetfulness, all in the vicinity, including innocents on the ground, were endangered. The piece concluded, "Women as pilots make good passengers" (Gould 1929, 691). Later, when more women became passengers and stewardesses became established figures, real anger became common.

In a sense, the publishing industry is responsible for the differentiation of male and female pilots, for the discussion of airwomen's presumed differences of motivation, and for the identification of distinct aviation roles. Newspapers and books reported the interest in fashion and family and chronicled the decision of airlines to hire stewardesses. The situation is, however, more complex than that. Newspapers, in particular, were often in the role of overtly supporting female pilots while reflecting a general set of social expectations that undermined their ultimate potential for success. Much of the press appears to have understood this dilemma and taken steps to blunt the worst effects. The vehicle was editorial comment, which often sounded warnings about the content of the news.

For example, as early as 1927 the *New York Times* bristled at the tendency to comment on female fliers' appearance, urging writers to get out of the "lipstick phase." It similarly disapproved of sympathy expressed for the husbands of pilots, usually of the "who will do the housework when they are off flying?" variety. Noting that male pilots are not subject to similar commentary, the paper expressed the hope that "when the novelty of a woman's piloting an airplane has worn off, she will no doubt be judged on a fairer basis" ("Fair Play for Women Fliers," 1927).

George Palmer Putnam himself wrote popular articles about being the husband of a pilot. Although he does comment on some of the frustrations of that role — being asked the same questions time and again by the press only to be misquoted, feeling like a tag-along — he hardly does this in a manner to elicit sympathy, to detract from Earhart's image, or to suggest a change in domestic relationships. On the contrary, the articles are designed to promote Earhart to the hilt. To Putnam, there is nothing at all wrong with being a "forgotten husband" (Putnam 1932, 1935). Of course, few others had such a strong financial stake in the issue, but the message comes across as generally applicable: pilot-wives are perfectly all right, says a noted New York publisher, who was not necessarily known by readers to be a behind-the-scenes manipulator.

There does not appear to have been much in the way of direct challenges to women's assignment to the peculiar roles of passenging and stewardessing, but there was objection to perceived inequality. The *Literary Digest* expressed disapproval that Helen Richey was consigned to the lesser tasks of stewardessing when she was not actually flying ("Queen Helen of the Air Breaks Down Industry's Last Barrier," 1935). Moreover, one finds considerable editorial insistence that men and women should have access to the same opportunities in piloting. In 1930, a New York feature writer proclaimed the end of the pioneering period of aviation. He noted that women had shared honors with men in this phase, showing "superb skill in piloting a plane." He asserted that on the basis of the rights so earned, women should expect full participation in the forthcoming "useful" phase of aviation, including careers as military pilots (Martyn 1930). A year later, the *New York Times* commented on the high competencies of female aviators. Apparently sensitive to suggestions that female pilots should somehow be thought of as different, the editorial dismissed any sort of segregation of airwomen as an outmoded idea: "All that has passed with their mastery of flying" ("Women in Aviation," 1931).

On September 6, 1936, Beryl Markham flew a tiny Percival Mew Gull from England to Nova Scotia, the first westbound transatlantic solo crossing by a woman pilot. This remarkable feat was well appreciated both in the United States and Europe. Indeed, Markham took up residence in the United States and was for a time something of a celebrity in southern California. In recent years, interest in Markham has been revived by the reissue of her book, *West With the Night* (1983), considered to be a literary masterpiece. Less well known is that within the same week, Dick Merrill completed a successful Atlantic crossing eastbound. Such heavy traffic occasioned some editorial commentary, most of it centering on the reliability of equipment and the implications for eventual scheduled transport. A particularly interesting editorial of the time juxtaposed the two flights, pronouncing Markham's the more significant on the basis of prevailing winds and aircraft limitations. The salient point is that the editorial did not consider — indeed, it did not even mention — Markham's gender ("More Ocean Flights," 1936). The implication is clear: men and women play the aviation game on the same footing and the same sort of performance may be expected of them.

Of course, it was relatively easy to argue for gender equity at an abstract level. It did not cost anything to advocate the cause of female pilots when the past or future was being discussed. The matter was slightly different when the issue was jobs, here and now. The cases of female professional pilots like Helen Richey, against the backdrop of the depression, and the WASPs, against the backdrop of war's economic dislocations, presented different challenges. Matters like these were a tougher test of support.

The press passed these tests, if not with flying colors. Richey, as noted in Chapter 1, was the first woman to fly for the scheduled airlines. After a brief tenure in 1935, she was pressured into resigning. This event attracted a great deal of attention and comment. A mini-symposium appeared in the *New York Times* ("Feminists Stirred Over Woman Flier," 1935). Featured was a scathing indictment of the pilots' union and the CAA by Amelia Earhart, an acknowledged heavy hitter. Her views were muted by the simultaneous publication of contrary opinions, particularly the more conservative position of Ruth Nichols, who doubted women's physical abilities to handle large transport planes. *Newsweek* likewise solicited input from a variety of sources in its feature article on Richey ("Women Pilots?" 1935). This article included a number of strong pro-Richey arguments.

Job-related hostility toward airwomen was an issue during these years, and the press recognized it. *Newsweek* seems to have consistently monitored employment opportunities for female pilots, for it commented on two important events of 1941. First, the CAA's Civilian Pilot Training Program, discussed in Chapter 1, banished women when it instituted the requirement that all graduates pledge to join the armed forces in the event of hostilities. This reversal of policy was publicly justified on the grounds that men made better pilots. Second, the magazine found some evidence of earlier Air Corps interest in female ferry pilots — probably something stemming from Cochran's early White House initiative or the conversations Love had with Olds at Ferry Command — and noted that nothing had been done to follow up. The editors, though not vitriolic in denunciation, clearly felt that there had been negligence, citing the successful British experience with female pilots of the Air Transport Auxiliary ("Feminine Flyers," 1941).

The disbanding of the WASPs was accomplished very quickly and at a time when the nation was absorbed by great issues of the war and its impending end. It therefore did not generate much in the way of commentary. The *New York Times* did not directly address the fate of the WASPs, but two notable editorials, one a year before and one at the time of the disbanding, spoke to the issue of female pilots as employed professional aviators. One commented favorably on the work of several female test pilots who had been employed by the Grumman Corporation. These women were not connected with the WASPs, but their function was very similar in that they flew first-line naval aircraft as part of the war effort. The paper lauded the company as well as the pilots and confidently predicted that professional female pilots were here to stay: "Test piloting is further proof that women have the poise and swift reactions necessary to handle the tremendous potentials of modern aircraft. . . . In peaceful days to come, they will be found among the most skilful users of private and commercial aircraft" ("Women Test Pilots," 1943).

The second, which appeared ten days prior to the formal decommissioning of the WASPs, addressed "Domestic Aviation." The promise of peacetime flying was the theme, but the editorial predicted that women would serve in the same piloting role as men. Without condemning the demise of the WASPs, it pointed out that the success of that organization and the vigor of its members virtually assured that many female pilots would make great professional contributions ("Domestic Aviation," 1944).

The premier, high-circulation general aviation magazine published an article that called the demise of the WASPs what it really was: blatant discrimination on the basis of gender. *Flying*'s use of terms like "blind, unreasoning sex prejudice," was strong condemnation indeed, and it followed a series of articles on the WASPs, its precursors, and the ATA, all of which were very supportive (Poole 1944).

Overall, readers after 1927 would have been aware that female pilots were in some quarters considered different from male pilots in motivation and basic nature. They were portrayed as fashion and family oriented. New, distinct, and inferior roles for women in aviation were thrust into public consciousness. While such developments seem ominous, the general press as well as aviation publications for the most part rejected these developments and continued to offer explicit support for individual airwomen and for an expanding, and gender-equal, professional role.

Airwomen on Film

Movies of the twenties, thirties, and forties dealt almost exclusively with stories, flights of fancy, or fictionalized accounts. They were thus very different in basic function from printed publications that disseminated information, editorialized, or served as a vehicle for biography. Nonetheless, the cinema was an important carrier of American culture, particularly during the later decades of the airwomen's era. Though it was an entertainment medium, it structured values and set agendas.

If the airwomen really sparked the public imagination, we would expect to find signs of it in the films of this era. While no real female pilot appears to have been the subject of a feature film, fictional heroines of the sky flew across the screen in profusion. Like the image of female pilots presented in newspapers and books, their portrayal was generally favorable. Moviegoers were frequently given the message that female fliers were both capable and courageous (Pendo 1985).

This is not to say, of course, that film producers unanimously used aviation story lines as a vehicle for some sort of comprehensive attack on prevailing mores about gender. Several early aviation films depicted very traditional gender roles. Howard Hughes's famous aviation classic *Hell's Angels* (1930) starred, in a subservient role, none other than Jean Harlow, one of the original movie sex symbols. Many aviation films, such as RKO's *Born to Love* (1931), portrayed women as sex objects in a love triangle involving two virile male pilots. Alternately, women were cast in the role of dutiful wife and mother, as a foil to a daring male who braved the dangers of the skies; a good example is MGM's *Test Pilot* (1938).

On the other hand, there were at least thirteen major U.S. films released between 1921 and 1943 that were quite different. They featured heroic and successful female pilots, some of whom were clearly modeled after the airwomen. In *Stranger Than Fiction* (1921), Katherine McDonald in her own plane boldly pursues fleeing aerial criminals and then extracts Frank Clarke, who was only partially successful in his tussle with the bad guys, from a seemingly hopeless situation. As a pilot in *Christopher Strong* (1933), Katharine Hepburn not only sets a round-the-world flight record but chooses suicide rather than retreat to a domestic lifestyle. Such a statement, especially when delivered so bluntly, had to be shocking in 1933. In MGM's *Too Hot to Handle* (1938), Myrna Loy plays a pilot who sets records and tests modern new aircraft. In another gender-role reversal, she engineers a dramatic aerial rescue of her brother.

These were notable films that featured some of the best-known stars of the day. Their potential social implications did not go unnoticed. Film historian James H. Farmer argues that screen portrayals of the aviatrix reflected "glimmers of change on many levels of contemporary society"

(1984, 88–89). As in the print media, this image was beclouded by the portrayal of new aviation roles explicitly for women. Simultaneous with the change in the press, films about airline stewardesses began to emerge in the mid-thirties. The genre continued over time and survives in contemporary porno flicks. Stephen Pendo (1985) identifies three particular examples from the day: *Flying Hostesses* (1936), *Love Takes Flight* (1937), and *Flight Angels* (1940). As one might expect, these films feature romantic or sexual relations between stewardesses and dominant male pilots. *Love Takes Flight* heightens the insult by portraying the failure of a stewardess who attempts to escape her subservient role. Beatrice Roberts rejects the job of serving passengers in the airliner cabin to undertake her own piloting venture, a solo transpacific jaunt. She falters in flight, but is saved by a man who stowed away on her plane. The film's message to any woman who aspired to the cockpit was, "Beware! You can be safe only if men are available to take over when things get tough."

There were World War II movies, such as *Flight for Freedom* (1943), that featured female pilots in heroic, quasi-military roles. But others expressed doubts about women's capabilities. The only major film about the Women Airforce Service Pilots, *Ladies Courageous* (1944), was nominally supportive of the patriotic service of the WASPs, but nonetheless capitalized on traditional and unflattering myths about women. Hysterical outbursts in the face of danger, competition for the attention of men, foolish flying errors, and suicide attempts are the order of the day. While contemporary reviews were negative and today's commentators regard the film as unfortunate, *Ladies Courageous* seems to have emerged from the same cauldron as the job-related hostility to the WASPs that was expressed elsewhere.

In sum, the female pilots' flight across the silver screen often verged on the spectacular. There can be little doubt that film contributed to a developing bright image. Movies of this period were fully consistent with the prevailing optimistic view of the future of women in aviation. But like the ominous signs visible in the press, tarnish around the edges of the silver could not be ignored.

Airwomen and Materials for Young People

Historically, defenders of the established social order have considered it crucial to get their messages to the younger generation — as have political visionaries and other advocates of change. For example, all modern political regimes, whether totalitarian or democratic, have sought to educate, indoctrinate, or socialize children into patriotic little citizens, while the formation of youth groups is never far behind the emergence of a revolutionary movement.

Prophets of the aerial age and aviation boosters of the twenties, thirties, and forties took strong measures to instill airmindedness in the youth of the United States. They sponsored aviation clubs, promoted model airplane competitions, modified the school curriculum, and with World War II on the horizon, formed paramilitary aviation cadet corps. Girls made up a significant fraction of the participants in at least the first two activities. In particular, girls were rather heavily into model airplane building; there were a number of clubs with exclusively female membership (Corn 1983, 117). Amelia Earhart tried to encourage this trend by offering special trophies for the best female entrants in national model airplane contests. Race pilot Gladys O'Donnell served as an adviser to a national youth aviation club that had many chapters.

The newsletters published by youth clubs and similar materials produced by early aviation boosters were fairly egalitarian with regard to gender. A charming cartoon from the period shows a boy who, attempting to assert masculine knowhow, is ultimately outclassed in umbrella "parachute jumping" by his sister (Corn 1983, figure between p. 70 and p. 71). Even so prominent a figure as World War I hero Eddie Rickenbacker got involved in the preparation of the young for the aerial age. He produced a long-running, nationally syndicated comic strip, "Hall of Fame of the Air." It featured exploits of a variety of heroic aviators, including airwomen (see illustration, p. 108).

Female pilots appeared in young people's fiction at a very early date. In 1911 Margaret Burnham began the *Girl Aviators* series (1911, 1912, 1913, 1914). These books, in print throughout the twenties, were typical of the era. The characters are courageous, well scrubbed, and though

Fay Gillis Wells was featured in the regular comic strip, "Hall of Fame of the Air," 1939. Smithsonian Institution Photo No. 84-18004.

perhaps a bit mischievous, thoroughly committed to the major values of the culture. The heroes uphold justice and goodness; the villains are unambiguously bad. The stories always end with evil punished and hard work and integrity vindicated. In short, the Girl Aviators, like the Rover Boys and Tom Swift, expressed a naive kind of adventurism and simplistic moralism that were the dominant forms of the age.

Burnham's heroes are three young women who live in far-off, mysterious Nevada. Each owns her own airplane, as do some young men who act as foils. The plots center around dangerous adventures, such as a cross-country trip from Nevada to North Carolina, or the courage of the central aviator, Peggy, who performs such feats as capturing a stage robber who had boarded a fast train following his dastardly deed.

The story lines are formulaic and hence not particularly notable. The portrayal of women, however, is of great interest. Gender stereotyping is almost completely absent from these books; indeed, the girl aviators are represented as better pilots than their male counterparts. Most of the characters in the books accept this as a matter of course. Occasionally an older person, typically a parent or uncle, expresses some reservations about women engaging in such vigorous activity as flying. This point of view never triumphs, however, and in fact is always shown to be anachronistic, not applicable to the new and different field of aviation. An especially aggressive assertion of women's role in aviation appeared in series advertisements at the back of the later books. "Aviation is not confined to the sterner sex as has been shown by Harriet Quimby and other daring young women. Girls who are fond of adventure will thoroughly enjoy reading these books which are wholesome and free of sensationalism" (Burnham 1913, 1914).

Similar themes were developed in other books right up through World War II. Many, such as Edith Lowell's Linda Carlton series (1931, 1931a), were astonishingly feminist. Linda Carlton not only learns to fly, but makes daring flights to summon doctors for the sick and beats male pilots in air races. Most interesting, Linda Carlton rejects the role of social debutante, giving up her coming-out party in favor of earning an air transport rating. She also declines a proposal of marriage in order to prepare for a New York–Paris flight, which she completes faster than Lindbergh. While Linda was represented as an outstanding and accomplished person, she was not portrayed as someone bizarrely unusual; other female pilots figure prominently in the plots.

To be sure, on occasion a fictional female flyer, though truly heroic, was made to depend on a man with whom she was romantically involved (Wayne 1933). Nonetheless, tough, independent female pilots were more common in young people's novels throughout the thirties and into the forties. For example, Bess Moyer (1932) created the Mapes twins, who are pilots for their father's business firm. Their most dramatic adventure is a flight to South America to find an entrepreneur who can save their father's financially threatened airfield. The trustworthiness of female pilots is a prominent theme.

Among the last of the genre was Margaret Irwin Simmons's *Sally Wins Her Wings* (1943). Sally appears at the outset to be a somewhat irresponsible type. She loses her job because she spends too much time at the airport and she is always broke — whatever money she has goes for aircraft rentals. But Sally persists, gaining more and more flying experience through a series of piloting jobs. Ultimately, she is accepted into the British Air Transport Auxiliary where she enjoys a wonderful career. The trials and tribulations on the way to success are formidable, and Sally is provided with an adviser and confidant. Notwithstanding the technical nature of Sally's interests, this person is another woman. Although exciting adventure is at its core, this volume differs from the others in that it has a pronounced career dimension; it was explicitly designed to be inspirational. To quote the dust jacket, "Sally Barnes is a typical American girl, ambitious, interested in a career, and willing to make great sacrifices to make a place for herself in her chosen field."

These novels replicate a theme that appears in other publications of the airwomen's era: it is not only acceptable, but desirable for women to be involved in aviation as pilots. The stories taught girls that women can aspire to professional careers as aviators and they can compete successfully with men. Implicit in these novels is the message that, though this may not be the case for old, established fields, there are no gender-specific roles in the new aviation enterprises.

Though these messages were visible in novels published from 1912 to 1944, a secondary development occurred in the early thirties. This development also had its parallel in materials discussed earlier. While it was young, the field of aviation had no distinctive role for women. With the advent of the airline stewardess, however, such a role emerged and was reflected in newspapers, books, and films. Young people's novels closely followed suit; after all, the target audience was at an age when career options enter consciousness. Ruthe S. Wheeler's *Jane, Stewardess of the Airlines* appeared in 1934, followed by Dixie Willson's *Hostess of the Skyways* in 1941, and Elisabeth Hubbard Lansing's *Sky Service* in 1942. The plots are predictable: fearful dangers are confronted and conquered and various bad guys are detected and foiled. In each of these books, however, the stewardesses adopt a subservient role. Helpful they may be; leaders or persons who use their abilities to the fullest they are not.

Stewardessing was difficult to represent as a continuing saga and so a series of books that featured the same hero was rare indeed. Nonetheless, Patricia O'Malley invented Carol, a star of multiple volumes. Carol's stewardess career, insofar as adventures were concerned, quickly reached a dead end and so it was necessary to move her out of the airliner cabin.

Piloting was apparently unthinkable to the author, for Carol became an assistant to a corporate vice president (O'Malley 1943) and an operative in an airline publicity department (O'Malley 1944). Compared to Linda Carlton, Carol's ambitions were restrained indeed.

Children's nonfiction featuring female pilots was quite rare, but some fairly powerful literature for girls appeared in the forties. Dickey Meyer's *Needed: Women in Aviation* (1942) has a heavy tone of wartime propaganda, urging vigorous patriotic participation. But her book was far more than a plea to come forward temporarily for the duration of the emergency, nor did it outline subservient roles for women: it envisioned female professional pilots in permanent key roles throughout aviation. Like so many other writers, Meyer accepted "the winged gospel" along with its glowing prophecy of the peace to come.

Jean Adams and Margaret Kimball's *Heroines of the Sky* (1942) is a compilation of short biographies of most of the major airwomen. It seems to be the first of its genre and may therefore have been of interest beyond its intended secondary school audience. These authors forthrightly encouraged girls to pursue careers as professional pilots and they reinterpreted the past to promote that goal. They wanted to affirm the abilities of airwomen who they believed had not been given due credit in school books and in aviation history. The book was remarkably candid. It pointed out that the airwomen faced great difficulties in an essentially sexist culture and warned its preprofessional audience that it could expect, at least in the immediate future, much of the same. In urging girls toward greater independence, it also cautioned against a condition that characterized most of the airwomen: "Most of the women who were able to get the essential tools for success in aviation had married pilots or promoters or men of wealth" (Adams and Kimball 1942, 293). Though the future female pilot will face difficulties, the gratification is worth it, these authors believed. They, like many others at the time, thought that World War II would severely erode traditional gender stereotyping. Adams and Kimball saw bright promise for the next generation and they boldly proclaimed it.

Conclusion

The airwomen, then, were well appreciated in their own time. By no means were they neglected heroes whose contributions can be appreciated only by observers who have the vantage of time. In Chapter 1, we saw that the airwomen's accomplishments were truly impressive — perhaps enough to have a long-term impact on the role of women in this culture. This chapter has demonstrated that the mass media perceived those accomplishments and portrayed them favorably. By and large, the media supported the airwomen and transmitted both information and positive evaluations to national audiences. Many of the airwomen wanted a more gender-equal society and they tried to exercise leadership through writing, often in popular outlets. The general press covered the airwomen generously; it was highly supportive even when they were under attack, and often enthusiastic about the future of women in the air. The new medium of the cinema sounded a nearly identical note and some of the biggest stars of the day played airwoman-like roles. Even the young were exposed to a matrix of literary support as both fictional heroes and, to a lesser degree, career advice endorsed the idea of women in the air. The combination, it might be thought, would profoundly affect the immediate behavior of contemporaries as well as lasting cultural norms.

As in Chapter 1, however, a very favorable general picture has some unsightly blots. Indeed, the mass-mediated word displayed some truly ominous developments. These are visible not so much in the degree of support that the press gave to the airwomen, as in evidence of cultural norms that the press reported. During the first quarter-century of flight, no particular activities were identified as uniquely feminine. With relative impunity, women occupied the same positions as men. After about 1927, however, female pilots were increasingly seen as different in motivation or orientation. They were often discussed in connection with fashion and family, while men's clothing or domestic situations never made the news. A traditional congeries of gender-based expectations began to creep into media discussions of aviation. At about the same time, and certainly related, some distinctive aviation roles for women began to take shape, the most visible and pernicious of which was the stewardess.

If women were supposed to think differently from men and do different things, why were they piloting airplanes? If airminded women could occupy the convenient pigeonhole of stewardess, what business did they have in the cockpit? Airwomen's actual behavior, however, belied differences; most demonstrated no intention of occupying new, subservient roles. Thus, in the minds of some, the airwomen were not doing what they were supposed to do. For the first time, the fact of women's commendable accomplishment was overlaid with threat and hostility. The media reflected these developments, but they did not, in general, indicate approval. The older, more egalitarian thrust remained strong and many magazine and newspaper commentators attacked the idea of special aviation roles for women. Unfortunately, this idea proved formidable. Contrary to what we might have expected, the pen proved unable to vanquish it. To proclaim the promise was not enough.

✿ 3 ✿

The Decade After World War II: Promise Denied

As we saw in Chapter 1, the development of aviation was a truly revolutionary event. There had been nothing like it before in human history and the public reacted with awe and wonder. Although not all of the grand benefits of an anticipated new aviation age came to pass, enthusiastic futurists were not entirely wrong. Profound social and economic consequences did follow the maturing of aviation. Naive expectations — for example, that aviation somehow portended a new form of prosperous egalitarianism — were doomed to disappointment, but it was certainly true that new opportunities were created. The airwomen, perhaps less constrained by dominant social traditions than established elites, aggressively took advantage of these opportunities.

Developments in aviation were in many ways a microcosm of a profound disruption in the general culture that occurred at the same time. The teens, twenties, thirties, and forties, though by no means revolutionary in terms of gender relationships, marked a period of widespread social change. Progressivism, Prohibition, the Great Depression, and the New Deal followed one another in close succession as the nation tried to redefine itself for a new era. All of these events in one way or another boded well for women; significant social disarray, after all, is most likely to benefit those who have traditionally been the most restricted. The overall social context within which the airwomen lived seemed conducive to at least modest change. It is certainly reasonable to think that this motivated, energetic, innovative group might have become great leaders in their gender's move into history's fast lane.

Economic distress and domestic political upheavals were not, however, the most dramatic historic events during this period, nor those of greatest consequence for women. Those honors surely go to America's great wars. War certainly has been one of the most powerful forces of social change; both "winners" and "losers" in modern armed conflicts are likely to emerge radically transformed. An international war, particularly a losing one, can unravel national leadership and lead to revolution, as in Russia in 1917. In a less violent but perhaps equally significant scenario, victorious nations may discover that the masses of people whose efforts were necessary for success are unwilling to retreat to subservience, as in Great Britain in 1945.

In a similar fashion, international conflict has profoundly affected the status of twentieth-century American women. The economic demands and opportunities of World War I brought unprecedented numbers of women into the industrial work force. Exposure to new experiences and new environments had a powerful impact on millions of women and created a new self-conscious awareness. Women liked being valued participants in a national effort and they became sensitive to the possibilities of political organization. The results were increased social activism, which ultimately led to the passage of the women's suffrage amendment in 1919, and the social emancipation of the 1920s best represented by the emergence of "flappers." Although the economic and professional gains made by women during the postwar years were smaller than some had expected and others had feared (Hummer 1979), there is no doubt that some symbolic barriers fell with a resounding crash.

During World War II, the mobilization of women for the defense effort affected all ages and classes. Compared to World War I, it was massive in size and scope. Not only did women enter the industrial and service economies in huge numbers, but they also joined the armed forces. This latter was a big departure from tradition: unlike many other Western countries, the American military had been almost exclusively a male preserve. This conflict brought forth "Rosie the Riveter" to work in factories, uniformed enlisted personnel and officers, and, as we have seen, women as pilots of the era's most sophisticated military airplanes. If the consequences of women's involvement in World War I were clearly visible, those following World War II could be expected to be nothing short of monumental. Certainly, some of the confident newspaper predictions about the golden future awaiting the postwar airwomen sprang from this line of reasoning.

Huge numbers of American women, at least nineteen million, were in the work force during World War II. This conflict was, in an unprecedented way, a battle of production: those nations that could deliver armaments and other war material in the greatest quantities were likely to win. This was fully understood by most protagonists from the beginning. Indeed, those Japanese military planners who opposed going to war foresaw that temporary fighting superiority would eventually be eroded through enormous U.S. productive capacity.

As the United States geared up its war economy, an acute labor shortage developed. Many women, behaving in text-book economic fashion, took the new job opportunities. But in addition, many went to the factory in response to a call to patriotic duty. Concerted media campaigns are credited with recruiting significant amounts of womanpower in major manufacturing centers like Detroit and Seattle (Rupp 1978, 74–114; Anderson 1981, 23–65).

Although the meaning and implications of women's employment during the war are still the subjects of much debate, it is clear that, in general, female workers were very successful. To the surprise of traditionalists, the jobs got done and productivity was high. But there were major social consequences as well as purely economic ones. Women themselves were changed. They entered a huge array of jobs never before available to them. After the war, women could envision a vastly greater scope of opportunity. Many, particularly those from the middle class, developed a kind of professionalism. That is, perhaps for the first time women enjoyed work outside the home; they undertook it for the sense of accomplishment and worth that it conferred rather than for remuneration alone. While working women have been part of the national economy for two centuries, this was a totally new type of female worker.

Employers were also changed. To some degree, they were forced to accommodate their new workers and, in the process, discovered that oldtime industrial management techniques did not always work best.

Command and submission, motivation through fear of sanction, and management-worker hostility had been the norm for centuries. Women did not respond as well to this model, in part because they were less often sole breadwinners and hence slightly less dependent on their jobs. Business managers were obliged to develop more "humanized" workplaces and these became, regardless of the gender of the employees concerned, the wave of the future (Weatherford 1990, 177–195).

Although World War II working women did not revolutionize prevailing attitudes about gender, they did gain a modicum of status and economic independence. Moreover, to a greater degree than many had expected, they remained permanent fixtures of the postwar U.S. economy. Many planners first thought of women's large-scale wartime employment as a temporary, emergency measure. It was regarded as abnormal and undesirable, inimical to family values and justified only by the extremity of the emergency. Most wartime female employees, at least initially, may have shared this view. They intended to work only for the duration of the war. But many came to view their working lives as promising and exciting, quite apart from financial gain and patriotic duty. A reluctance to surrender these gains was certainly in evidence by 1944. Female workers wanted to stay on.

This became a source of great consternation among political office holders, journalists, labor union leaders, and employers. In the mid-forties, to an even greater degree than today, Americans were fixated on jobs. The extent to which the economy was capable of expanding and the importance of productivity to prosperity were not clearly understood. This was especially true of the public at large, but highly placed business executives also seemed to view the world in terms of fundamental scarcity. It was assumed that in peacetime there were only a limited number of available jobs; how these might be parcelled out was a matter of critical concern. Many feared that there would not be enough jobs to go around if women did not relinquish their wartime employment once hostilities ended. *Atlantic* wondered how to "get rid of the women," while *Woman's Home Companion* urged its readers to "give back the jobs" to men (Weatherford 1990, 307). At the same time, as if to expunge the horror of the war from conscious attention, there was an attempt to recreate the society that had existed before. Such a society would presumably not incorporate those changes in the status of women that had come about during the conflict.

Heated debates ensued, similar in some measure to those that surrounded the disbanding of the WASPs. While many postwar women strongly claimed the right to stay in the workforce, others passionately desired to return to an idealized, prewar "normalcy." Still others wavered and waffled. The media made the same arguments, sometimes inconsistently urging both at the same time.

Ultimately, work won. To be sure, postwar demobilization resulted in four million fewer female workers, but a large proportion, perhaps two-thirds, of those who wanted to continue working were able to do so. Moreover, following the immediate shock of conversion to peacetime production, the proportion of women in the labor force continually increased. Men were not put out on the street; indeed, the number of employment opportunities expanded dramatically and unimagined prosperity developed. The American economy, far more robust than almost anyone had imagined, was totally transformed (Anderson 1981, 154–178).

The decade following the war thus might have seemed ideal for the airwomen. For thirty years they had demonstrated competence and, because they operated in a period of ferment and social disruption, those demonstrations had found fertile soil; others had been inspired and impressed. Now one might imagine that the women in aviation would move to a new stage: professional pilot positions. Many of the airwomen had laid out their professional aspirations in the thirties or before. At last, it seemed, their sound tradition and favorable economic conditions guaranteed success.

Nothing of the sort developed. The airwomen were not part of the emerging postwar future. Not only did progression to a new stage fail, but in fact there was a great backslide. Female pilots did not gradually fade from the scene; suddenly, they vanished almost without a trace. The contrast between hopes born of accomplishment and the predictions of confident editorials on the one hand and cruel reality on the other could hardly have been greater. The decade following the war was one of brilliant and unprecedented opportunity for many Americans, including many women, but for female pilots it was a grey period indeed.

The Disappearance of the Great Airwomen

Airwomen were thriving in the thirties. Some of the most prominent were excellent writers and many seemed poised for long-term leadership roles. Most were quite young, presumably looking forward to many years of productive career. World War II, terrible though it was, might have been a great positive stimulus to such people. If women in aviation were about to begin a new stage of professional development, however, the airwomen did not come forth to launch it.

To be sure, the most visible airwoman, Amelia Earhart, never returned from her 1937 around-the-world flight. Harriet Quimby, who had great writing skills, had died in 1912. Still, the great majority of the other accomplished airwomen not only survived, but in 1944 were entering the period of life that is often the most productive. Ruth Law had twenty-six years to live; Katherine Stinson, thirty-three; Marjorie Stinson, thirty-one; Phoebe Omlie, thirty-one; Louise Thaden, thirty-five; Ruth Nichols, sixteen. Elinor Smith, very much alive as of this writing, may well log a half century beyond this critical date. What did these people do following World War II? The answer, it seems, is very little. In actuality, many had withdrawn before hostilities started, and whatever the years between 1941 and 1945 may have meant to them, they did not reemerge. Their story is one of potential unrealized, abilities unused.

Ruth Law was a person who was not easily discouraged. In fact, her career reveals one very dramatic comeback. Her request to become a combat pilot in World War I was denied, but this did not deter her from an aviation career. A postwar trip to the Far East led to many firsts and Law resumed an extensive schedule of profitable exhibition flying. In short, she began exactly where she had left off before the war. Reportedly, she could earn as much as $1,500 per day, a veritable fortune in this era.

In 1922, however, Law discovered, by reading the newspaper, that her flying days were over. Her husband, without her knowledge, announced her retirement to the press. He further told her, "The statement is bona fide. I can't stand the strain of seeing you in danger. You've tempted luck long enough. To please me, give it up." Law's reaction is worth quoting at length:

What else could I do? He had been my backer, my manager and my staunchest friend. I, too, realized that perhaps the strain of watching had been greater than the strain of . . . flying. . . . I stopped, but I stopped too suddenly. . . . I kept my nerves under control for two years and then I had a nervous breakdown. Just the sight of good looking aviation clothes makes me homesick for the flying fields. Sometimes the sight and sound of planes overhead is maddening, but my flying days are over. I sold everything I had that pertained to flying. My only keepsakes are a propeller . . . and the map I used on the Chicago to New York flight (Gee 1932).

In 1924 Law expressed some interest in a transatlantic flight, but nothing ever came of it. She spent her time in the somewhat less demanding pursuit of rock gardening. Law rarely spoke publicly after that, and she never flew a plane in the last forty-eight years of her life.

The Stinson sisters similarly withdrew from aviation. Katherine's post–World War I career was brilliant but depressingly brief. As noted in Chapter 1, she toured widely as an exhibition pilot and, through persistent demonstration of piloting skill, in 1918 secured an appointment in the U.S. Airmail Service. This was potentially a very important precedent. The airmail service was new and its pilots were among the first true aviation professionals. Regrettably, Katherine resigned after only a few flights and so a potential role model was lost. It is possible that this decision was motivated in part by poor health, but it is certain that Katherine married a judge and settled down to relative obscurity in a remote New Mexico town. She certainly did not exhibit any leadership in the field of aviation; in fact, there is no mention of her in the *New York Times* between 1919 and 1977, the year her obituary appeared.

Marjorie Stinson remained an active pilot, as evidenced by the fact that in 1929 she became a charter member of the Ninety-Nines, an organization of female aviators that persists to this day. Although she did pursue a career in aeronautical drafting, she was not an aviation leader. Between 1919 and her death in 1975, the *New York Times* considered her newsworthy only once, when she was a passenger on a flight over the Panama Canal.

Phoebe Omlie was an innovative leader in the Air Marking Program of the thirties. She followed this with work for the Tennessee Aviation

Commission and, during World War II, an appointment with the Civil Aeronautics Authority involving site selection for ground-school training (Buffington 1968). After the war Omlie did nothing to resume a prominent role. Much of her energy seems to have been absorbed by a strident attack on the state of public education, though few of her insights in this area have survived in print.

We have already examined the off-again–on-again character of Louise Thaden's career. By 1937 she had essentially given up aviation, though she did write her thought-provoking autobiography in the following year. Thaden remained an active pilot well into the seventies and later in life spoke at certain aviation events, but she never made the news after 1937.

Elinor Smith prepared her autobiography in 1981. She was the last of the great 1930s airwomen to do so. This book is remarkable in that the narrative ends virtually fifty years before the date of publication. Smith does not tell us about the last half-century of her life, but the clear implication is conventional familism. There is no indication of any involvement with aviation. Perhaps Smith's feisty character made it difficult to stay prominent in the community of female aviators. She certainly did not shy away from criticizing her peers; Amelia Earhart was often a target of barbs. In any event, Smith had faded from the scene by the forties and was virtually out of public view until the belated appearance of her book.

Colorful, energetic, and perhaps foolhardy, Ruth Nichols enjoyed a dramatic career throughout the thirties. Although she did not appear very active during World War II, Nichols tried after 1945 to reestablish herself as a leader in the use of aircraft for humanitarian purposes. She was involved with an organization called "World Flight" as well as with UNICEF. Indeed, it was on a United Nations junket that Nichols experienced yet another serious airplane crash — to add to her existing string of six. In 1949, she was aboard a four-engine C-54 returning from Europe when, through a gross miscalculation, it ran out of fuel and was ditched in the Atlantic. Incredibly, she and all but eight of those aboard were dramatically rescued at sea (Nichols 1957).

Nonetheless, aviation and humanitarianism did not, during these years, appear to be a particularly good mix. Perhaps Nichols clung too long to an overly optimistic and outmoded notion from "the winged gospel."

Her impact was decidedly limited. In postwar years, she made the news only twice: once in regard to the Atlantic mishap, and once in 1957, when she was appointed to a position with the National Association for Retarded Children, which had nothing to do with aviation. As if to symbolize despair among all the airwomen, Nichols died in 1960, an apparent suicide ("Death Ruled Suicide," 1960). The proud airwomen of the 1930s had sunk to a depressing new low.

The Military Airwomen: The Aftermath of World War II

A cynic, or even a reasonably neutral observer, might have thought that World War II would snuff out the last vestiges of public optimism about aviation. A naive hope of a bright future based on flight had characterized the American public for decades. In 1945, however, any promises that had been realized were juxtaposed against a grim military reality of previously unimaginable scope. Ironically, some of the war's greatest horrors — the deaths of millions of civilians by bombing, the introduction of atomic warfare — were now part of aviation's legacy.

Despite the association of airplanes with indiscriminate carnage and mass destruction, many regarded World War II as a mere temporary perturbation in the nation's aerial voyage to the promised land. They believed that all the great power of flight, like atomic energy, would be harnessed to humanitarian goals. The best example of such thinking is perhaps the many wartime newspaper features speculating on the role of aviation in the sunny days to come. Indeed, the subsequent peace was thought to signify the realization of all that was good about aviation. Expectations once again rose to giddy, unrealistic heights. Such was the power of "the winged gospel" that the worldwide horror of war could be treated not as a failure of prophecy, but as the precursor to vindication.

Nor was such thinking confined to an impressionable, emotional public or to journalists trying to create stories consistent with postwar euphoria. Hardheaded businessmen confidently thought that the war had stimulated a huge public appetite for flying; they fully expected that all those who had flown in the war would continue as civilian pilots. Factories all over the country started spewing out huge quantities of single-engine light airplanes for recreational or business use. Happy families enjoying aerial vacations and executives on a quick, winged commute to work were the subjects of sophisticated mass advertising campaigns. Color display ads for Cessnas and Pipers appeared in national general-circulation magazines like *Life* and *Saturday Evening Post,* right beside those extolling Plymouth, Frigidaire, and Electrolux.

This great aerial expectation was soon awash in disillusion. There were two great flaws in the prevailing corporate reasoning. First, fascination with aircraft and even wartime experience with them did not translate directly to a motivation to buy and fly. While "the winged gospel" had very powerful emotional and symbolic dimensions, the behavioral ones, it turned out, were less developed. Second, sheer economics intervened. The price of aircraft remained relatively high. Although large-scale production can drive unit costs down, airplanes require a great deal of precision in manufacture, considerably more than for autos or refrigerators. To the surprise of many, airplanes remained too expensive for all but the wealthy. Individual income levels did rise in this period, but not enough to bring the flying machine into the range of the masses. By 1949 the bubble had burst and production of general aviation aircraft retreated to much more modest levels; by the early fifties, many aircraft manufacturing firms no longer existed, including that founded by the Stinson family.

There were other disappointments. Derivative enterprises in training, maintenance, and facilities construction also failed to develop. Commercial aviation, though it was to become immensely significant in the decades ahead, got off to a slower start than many had anticipated. Again, things aerial remained quite expensive despite confident predictions to the contrary. While postwar booms were common throughout much of the American economy, aviation, at least at the popular level, was left out.

As a result, the large number of pilots who had flown for the military or military contractors did not find ready cockpit employment. At least temporarily, there was a very limited number of pilot positions available. The surplus of aviators that developed in 1944 and led to the disbanding of the WASPs continued into the postwar years.

The situation was bleak for many ex-military pilots, but especially so for women. Indeed, the postwar behavior of military airwomen was in many cases similar to that of women who had been involved in the war industries. Many went home and assumed traditional, familial lifestyles. But unlike the case for many others, this was a forced choice. It is certainly inaccurate to argue that they "returned" to such a domestic condition, for few could claim an early history of conventional gentility. Moreover, there is ample evidence that what ex-WASPs in fact did after 1944 stands in some contrast to what they wanted to do. Perhaps the best-known — and certainly the most dramatic — example involved the WASPs detachment at Wilmington, Delaware. On December 20, 1944, the very day of disbandment, Betty Gillies, squadron leader of the Second Ferrying Group, and all forty-two women under her command volunteered their piloting skills to the government for one dollar per year ("WASPs Volunteer Services to Army," Associated Press dispatch, Dec. 20, 1944). The Second Ferrying Group was not alone in making such an offer. WASPs who had served at Romulus, Michigan, sought positions as company pilots with Bell Aircraft in order to facilitate the delivery of P-63s, which were backing up at the factory door (personal communication with Elizabeth Boyd, April 16, 1992).

Although these women certainly wanted personal gratification and sought to advance their careers by such acts, they were not entirely self-serving gestures. The war was not yet over and there were still many airplanes to be ferried. Many of the newly available male pilots were not experienced with advanced aircraft and would require significant training to be brought (literally) up to speed. General Arnold denied all of these requests: "You will notify all concerned that there will be no — repeat — no women pilots in any capacity in the Air Force after December 20 except Jacqueline Cochran. I do not want any misunderstanding about this so notify all concerned at once" (reprinted in Granger 1991, 469).

This was not the only kind of rejection the airwomen received. Though the WASPs formed under military auspices, the pilots, as noted earlier, remained civilians; they were not part of the Army Air Corps nor of the Women's Army Corps. Many people believed that simple equity demanded some sort of militarization, if for no other reason than pay, but this did not happen. In fact, in 1944 a bill to provide the WASPs with true military status and hence veteran's benefits, a matter of increasing concern as the end of hostilities loomed, failed in Congress. Coupled with demobilization and banishment from military aircraft, this was a true slap in the

Women Airforce Service Pilot H. Richards in an open cockpit trainer, c. 1943. U.S. Air Force Photo Collection (USAF Negative No. 118315 AC). Courtesy National Air and Space Museum.

face. Many of the airwomen felt confusion and shame and were unable to understand what was happening to them (Granger 1991, 473–480). Many WASPs simply went home defeated, bitter, and wondering why they were no longer valued.

Betty Gillies did secure a position as test pilot for Ryan Aeronautical, but it did not last long. Moreover, it is unlikely that her appointment signified anything about the general willingness of the U.S. aviation industry to hire women; Gillies's husband was in charge of certain new product development programs within that company. In 1947 three ex-WASPs landed an assignment delivering a group of Beech Bonanza airplanes to Brazil. A few WASPs took jobs with aircraft manufacturers, dealers, or fixed-base operators. These tended to be "utility" occupations, which included some delivery and demonstration piloting. At least one returning ATA girl was hired as an instructor. Some of the economically more fortunate airwomen purchased light planes — often in cooperation with husbands — and remained general aviation pilots (Douglas 1991, 58–60; Verges 1991, 224–229).

The postwar career (or lack thereof) of Nancy Harkness Love is in many ways typical. As we have already seen, her war record was nothing less than spectacular. She was a key figure in establishing and organizing the WAFS. Despite the awkward position of this new function in the Army chain of command and in the Washington bureaucracy, and despite bitter squabbles over the way female military pilots should be organized, she effectively ran her unit throughout the war. Like many successful administrators, Love kept her hand in the actual substantive business of the organization. She continued to pilot airplanes throughout her tenure, delivering a four-engine C-54 transport on the very last day of the WASPs' operational existence. Astonishingly, she was just thirty years old when she wrote her last reports for Air Transport Command. Her demonstrated ability as a pilot, clear leadership aptitude, dramatic organizational skills, and inventive energy represented a tremendous potential resource. Moreover, there was plenty of time for the realization of that potential. Though Love would die of cancer at a young age, she had in 1944 thirty-two years remaining (Powers 1991).

Love, however, disappeared into domesticity. She bore three children while her husband, Robert, was involved in the founding of the corporation that would become Allegheny Airlines. The family later moved to Martha's Vineyard, where Robert was involved in boat building: "Nancy crewed for Bob on their sailboat, rode horseback with her daughters, and enjoyed an occasional visit from one of her WAFS. [They] limited their flying to taking the children off-island for doctor and dentist appointments" (Verges 1991, 229). Given this lifestyle — more appropriate to a retiree than a multitalented dynamo — it is not surprising that Love never made the news after 1944. Indeed, the New York Times did not even acknowledge her with an obituary.

Some observers have wondered why Love's withdrawal was so sudden and so complete (Douglas 1992, 53; personal communication with Deborah Douglas, Dec. 3, 1991). There is some evidence that she would have preferred a continuing leadership role in aviation, and apparently she

Nancy Harkness Love in full WASP uniform, c. 1944. U.S. Air Force Photo Collection (USAF Negative No. D-24844 AC). Courtesy National Air and Space Museum.

contemplated major initiatives in other areas, such as the founding of a
boarding school. None of these things took place. It would seem that a
kind of depression settled over Love, one with very long-term effects.
Whether this was in any way caused by the nation's unceremonious
rejection of women military pilots involves sheer speculation, but it is true
that Love's behavioral pattern was repeated among hundreds of other
WASPs.

Evidence of self-destructive depression is somewhat stronger in the
case of other military airwomen. Helen Richey pursued a professional
aviation career more aggressively than any other airwomen. She was in
fact the first female pilot on the scheduled airlines, served with the CAA's
Air Marking Program, and did some air racing. It is hardly surprising that
she would seize the first opportunity to fly for the military, and Richey in
1941 volunteered for the British Air Transport Auxiliary. This brilliant
array of experience made her one of the most notable American recruits.
Though she served well and the need for pilots in Britain continued to be
critical, Richey chose not to remain beyond her original contractual
commitment. Preferring to serve her own country directly, she returned to
join the WASPs, one of only three ATA girls to follow this pattern. Her
U.S. duties began in 1943. Her career, primarily as a ferry pilot, was
commendable; for example, she served a stint as commanding officer of
the WASP unit at Fairfax Field in Kansas City. However, following December 20, 1944, Richey's professional string ran out: there were no
military airwomen; there were no female pilots on the commercial airlines.
Two years later she was dead, a suicide.

Mary Wiggins had been a Hollywood stunt artist in the thirties and
a member of a depression-era flying club known as the Women's Air
Reserve. She joined the WASPs in 1942 and, like Richey, was completely
successful; indeed, she too served as a commanding officer, in this case at
Hondo Army Airfield in Texas. Following disbandment, Wiggins returned
to the movies and resumed her job as a "double" for famous actresses,
among them Barbara Stanwyck, Dorothy Lamour, and Claudette Colbert.
Though outwardly very successful, Wiggins was reportedly "disappointed
in everything." She too committed suicide shortly after the end of the war
("Veteran Woman Stunt Artist Ends Own Life," 1945).

Helen Richey in Air Transport Auxiliary uniform, c. 1942. International Women's Air and Space Museum Photo No. C 114.

Of course it would be irresponsible to infer a grisly pattern from these two events, but coupled with Nichols's demise they do suggest that despair and a diminished sense of self-worth were common among this group. Certainly it symbolizes in a profoundly shocking manner the end of the golden era of the airwomen.

Postwar Female Pilots

The great airwomen had prematurely taken their leave. The society had tried to banish the military airwomen to the scrap heap of history and in response most of this group also quietly disappeared. But were there not others — younger perhaps — ready to step into the breach? In business, politics, the military, and certainly most social movements understudies and heirs apparent are always waiting in the wings for the old guard to move on. New standard bearers often revitalize old programs and pour new energy into important causes. Indeed, at least from the longer-term viewpoint, the departure of the airwomen would not necessarily have been a tale of unrelieved tragedy and woe. Quite the contrary, it would be reasonable to see in it opportunities for renewal, for the infusion of new blood, for the development of new agendas and progression to a new stage — a badly needed professional stage — in the progress of women in aviation.

Indeed, women did not completely abandon the skies after 1944. Conscious attempts were made to reestablish women's air races and there were even a few attempts at record setting. Some women intended to pick up exactly where things had left off at the beginning of the war.

The Ninety-Nines, an organization of female pilots that had existed since 1929, suddenly had a larger pool from which to draw membership. Women had been trained as pilots during the war through the WASP and other defense-related operations. While professional opportunities for them were few, they nonetheless were present in some numbers. That enlarged organization quickly organized an all-woman airshow, which was to be an annual event featuring various kinds of races as well as aerobatic competitions. The first all-woman show was presented in 1947 at Tampa, Florida, drawing some thirteen thousand spectators. A key event at the

1948 show was the Cochran All-Woman Trophy Race, a coast-to-coast competition. A finish line was laid out at Tampa and each competitor's completion of her long journey was part of the excitement of the airshow. This competition, the purse for which was supplied by Jacqueline Cochran, evolved into the annual All-Woman Transcontinental Air Race (AW-TAR). Without doubt this was the best-known women's aviation event in the postwar years. It was dubbed the "Powder-Puff Derby," as was its precursor, the 1929 Women's Air Derby described earlier.

AWTAR was a true competition, but it was heavily constrained by rules and regulations. High-powered ex-military craft were deliberately excluded and an elaborate handicapping system made it possible for a great variety of aircraft to compete in a meaningful way. This was not a flashy event, but it did provide its participants with an opportunity for fellowship and social interaction. In a way, AWTAR was a symbol of the maturity that aviation had achieved in the thirties: sheer speed and absolute records were not the goal; practical application was. Important aviation figures regarded it as "the best single showcase of dependability of the light airplane and its equipment" (Douglas 1991, 74). AWTAR was held each year through 1977.

Though the ambience was very different from the prewar years, some female pilots rose to prominence. A few had roots in the past and credentials qualifying them as airwomen. Certainly among the most notable was Blanche Noyes, race pilot, veteran of the 1929 Air Derby, and participant in Omlie's Air Marking Program. She was an important figure in organizing the postwar shows and races, serving as president of the Ninety-Nines in 1948 and 1949. She remained as the director of the Air Marking Program and continued to urge communities to paint navigational symbols on the tops of their water towers.

Edna Gardner Whyte had won a number of closed-course races between 1934 and 1941, though she certainly did not achieve the level of national fame that many other airwomen did. During World War II she was an instrument instructor for the Spartan School of Aeronautics, which held a contract to train Army pilots. The cancellation of training starts in 1944 cut Whyte's flight training career short. Unlike many of her male counterparts, however, she did not respond by attacking the WASPs as usurpers of jobs; instead she resumed a former profession, joining the Army Nurse Corps for the duration.

After the war, Whyte was hired as a flight instructor by a Fort Worth entrepreneur who envisioned thousands of veterans using their GI Bill benefits for pilot training. Like most of the postwar female aviation professionals, she became something of a jack-of-all-trades, delivering and ferrying planes, among other tasks. Whyte found a number of supportive people in this environment and renewed her old passion for racing. She became one of the leading competitors in closed-course races. A few dramatic events for women were held until about 1949. The designated machine for these competitions was typically the T-6, a powerful advanced trainer available in quantity as postwar surplus. When cross-country events became the mode she took those on as well, though her participation in the AWTAR itself was marred by a disqualification controversy that apparently caused bad feelings for a number of years. Whyte gained greater fame later in life. She remained an active instructor and aviation entrepreneur well into her eighties, standing as a symbol of accomplishment not only for women but for elderly people as well (Baxter 1984, 108).

Grace Harris had been an active pilot in the 1930s, but had not been a competitor or otherwise drawn any public attention. During World War II she worked for a fixed-base operator and so had a real aviation role in the defense effort, though one that did not involve dramatic service as a pilot. Harris, like Whyte, found herself in a conducive environment following the war because many of her friends and colleagues were in air racing. Harris was clearly the most successful of the female T-6 racers, winning the major national event in 1948 and again in 1949. Harris had a great deal of energy and channelled it in many directions. She was heavily involved in sports car rallies and aerial travel. In addition, she compiled a great record of success as a businesswoman in various aviation-related ventures (Harris 1980).

Fran Bera competed in the AWTAR twenty-one times and placed in the top ten in seventeen of the races (Douglas 1991, 76). Marion Hart learned to fly in 1946 and later dramatically flew a light plane across the Atlantic (Hart 1953). Molly Bernheim similarly caught the brief wave of post–World War II aviation enthusiasm. As new pilots she and her husband flew extensively, providing the basis for an anecdotal chronicle not unlike that penned by Bessie Owen in 1941 (Bernheim 1959).

These women cannot, however, be called daughters of the airwomen. First, they seem to be quite different motivationally. Some, like Noyes, seem to have accepted the notion that women's roles in the postwar order were to be different from those of men. To be sure, Noyes's career clearly demonstrates that she did not believe women's place was in the home; she would not even have agreed that women's ideal aviation role was that of stewardess. Nonetheless, she demonstrated a strong urge for the postwar restoration of many traditional feminine values: "Noyes . . . had captured the essence of the unique experience of women pilots. This was maintaining a feminine appearance, planning activities independent of the major 'male' shows and races, and always behaving like 'ladies' " (Douglas 1991, 64). As we have seen, emphasis on the unique character of female pilots was common in the prewar years, and even some of the airwomen indulged in it. By the mid-forties, however, it was clear that from the standpoint of many women's needs, this was counterproductive. Certainly it was not consistent with the aggressive development of a new professional stage for women in aviation. Professional advance, then as now, depended on aggressive presentation of productive ability and competence on the one hand, and vigilance against discrimination on the other. A special role for women obscures the necessity of achieving high levels of competence and may even justify discrimination. Definition of a special role can, however, serve needs for fellowship and camaraderie. The Ninety-Nines, under Noyes, appeared to serve the membership by providing them with pleasant society, which had not been one of the airwomen's traditions.

Noyes compounded the error by a continued insistence on the importance of air marking ("Resumes Air Tours Urging Towns to Paint Names on Barn and Factory Roofs," 1947). As we have seen, the Air Marking Program had been conceived, planned, and implemented by airwomen. Indeed, it had become a women's preserve. While visual cues were indeed important to early cross-country air travel, they were far less so after the war, in an era of increasingly sophisticated electronic navigation devices. Noyes appears to have pushed the matter because of the traditional involvement of women, ignoring the fact that while it was a brilliant notion in its time, that time had passed.

Harris and Hart had radically different views from those of Noyes, but neither were they cut from the same cloth as the airwomen. These two were extremely impressed by their own success. They seemed to revel in that success to the extent of being pleased that few other women could match them. Harris and Hart, were, to be sure, exceptional women who went far beyond what traditional feminine stereotypes of the age dictated, but they did not see themselves as exemplars for other women. While they may have recognized that the deck was stacked against women, they felt no outrage at this; instead, they were smug in their possession of the skill that allowed them to overcome it. Instead of viewing other women as a group that could achieve great things if provided with leadership, Harris and Hart viewed those women with contempt.

In 1980 Harris wrote *West to the Sunrise,* a title that apparently derived from a trip from France to the United States aboard the Concorde, the world's only operational supersonic airliner. The Concorde's speed is greater than the earth's rotational velocity; travelling west, the Concorde eventually "catches up" with the sun. It is indicative of Harris's focus on status and privilege that she clearly regarded such a trip as her own just dessert. Harris was an extremely able woman who, bored by being a housewife, undertook a business and flying career. She was greatly pleased with her successes and justifiably so. Her 1948 and 1949 racing victories are the major topics of her autobiography. Like all major races, these were in large measure a team effort. Harris understood this and acknowledges the planners, the technicians, and the entrepreneurs. She does not, however, celebrate their collective accomplishment; rather, she is immensely gratified that all these skilled people are so devoted to her personally.

Her description of a European tour under the auspices of the FAI is in the same vein. So many people, especially an ex-Nazi Austrian military officer, were so willing to help her deal with these strange and vaguely unpleasant foreign conditions. Had she been a few years younger, Harris would have been well qualified as a member of the "me generation."

To be sure, at one point Harris does discuss the role of women in society, but this discussion is bizarre in the extreme. It begins innocently enough with the assertion that women have been very important to aviation. To make this general point, she discusses Jerri Cobb. Cobb made a concerted effort to become the first female astronaut, an effort Harris lauds. However, Harris misses the whole point of this incident: despite impressive qualifications, Cobb was not selected for the astronaut program. Harris is not indignant; she fails to understand that if this is the way competence and hard work are rewarded, few women would follow Cobb's example. To Harris, it is just somehow nice that Cobb put forth this great effort, but the end result is of no consequence. How different Harris is from the airwomen is best indicated in the foreword to *West to the Sunrise*. To be sure, it is written by a male, an Air Force general named Brooke E. Allen, but he was chosen specifically in order to sing her praises. Allen notes, "I am somewhat puzzled by the feminine movement today for equal rights. Success in any field is attained by ability and determination; sex is seldom a factor."

Marion Hart was equally uninterested in the potential social consequences of her flying career. The title of her book, *I Fly as I Please*, likewise betokens a radically self-centered approach. Her book is basically a travelogue by a person who was rich and condescended to describe the experiences that only one of her status could have. She was immensely proud of that status and clearly wanted as few others as possible to share it. Hart had disparaging remarks about social change generally, gender equity, and the Ninety-Nines, the conservative nature of that group notwithstanding.

Hart's lack of concern about other women in aviation fits her general message, but her complete alienation from them comes as a surprise. Her older sister, she tells us, learned to fly in 1916 — Hart thought this was a first for a woman. That she would be ignorant of Quimby, Law, and the Stinsons tells us she had little interest in the significance of her sister's achievement, and probably anyone else's.

Edna Gardner Whyte had, by 1938, accumulated flying hours equal to those clocked by Louise Thaden and Phoebe Omlie. It is thus not surprising that in *Rising Above It*, written with Ann L. Cooper in 1991, she presents an airwoman-like sensitivity to gender:

> I had lived for eighteen years before women were "given" the right to vote through ratification of the nineteenth amendment and I share the indignation with which feminists bristle at such a patriarchal inference. . . . I specifically chose to enter aviation, a man's field in a man's world spurred by a fierce desire to *be* someone. . . . I cheer my

female flight students when they aim for the captaincy of commercial airliners. Can I help but feel jealous that I was never allowed to be selected? I cannot help but resent the damnable men who were at aviation's helm and who stood in the way when I sought commercial flying jobs (Whyte and Cooper 1991, 1–2).

The sexist nature of society troubled and sometimes enraged Whyte at virtually every stage of her life, not only on the three occasions, in 1935, 1940, and 1941, when she vigorously sought employment as a commercial airline pilot. She deeply resented exclusion from races, limited business opportunities, and the sexist slurs of airport gossip. She well understood that her accomplishments might have been much greater and have come much sooner if she had not faced overt discrimination. Although this realization distinguishes Whyte from Harris and Hart, it does not seem to have led her to a particularly enlightened strategy. Her response to gender discrimination was self-aggrandizement through manipulativeness and, quite possibly, man-hating.

Whyte managed to get special favors from people who had what she needed; these were, of course, usually men. Often the special favors involved a degree of insult to the very men who had granted them. Just a few days after soloing, she borrowed a plane from her flight instructor, who was also her lover, in order to take another man on a weekend getaway. Of course, such an event is not in and of itself particularly shocking or even noteworthy; what is curious, however, is that Whyte would see fit to expound upon it in her autobiography.

Elsewhere, Whyte comments on long-time companion Guerdon Brocksom: "Brocksom let me call his [airplane] my own. He flew it from Newport to Washington and left it with me to fly and demonstrate. . . . I thoroughly enjoyed knowing that I had an airplane at my disposal. . . . No longer did I have to . . . connive an airplane from someone else." A few months later Whyte entered the airplane in a Florida race and was relieved to hear that Brocksom was unable to attend this race: "I hadn't told him that I'd met another man in Washington: Ray Kidd, a dark-haired, urbane writer for the United States Information Service. Kidd held promise of being *my* G. P. Putnam" (Whyte and Cooper 1991, 161–163). The baldness of such a statement is startling, but it is in keeping with sentiments Whyte expressed throughout much of her life. Whyte remarked of the period shortly before her divorce from second husband Murphy Whyte, "My lust for racing flamed as my marriage cooled, although I continued to depend on Murphy for race tactics and his superb mechanical abilities" (Whyte and Cooper 1991, 233).

Although Whyte was a member of the postwar Ninety-Nines, the idea of cooperating with other female pilots to defeat society's entrenched gender bias, to achieve common goals, and to affect the future was foreign to her. To be sure, this was by no means unique to Whyte. Indeed, some of the 1930s airwomen — Pancho Barnes comes to mind — thought in very similar ways. Nor is this mindset necessarily undesirable; Whyte was immensely successful and so her motivation must be regarded as productive. By 1944, however, this extreme individualism was somewhat out of date. It was consistent, perhaps, with the old demonstration stage of women in aviation but it had limited relevance for a new professional stage. Indeed, Whyte viewed her postwar racing career as a personal extension of the pre–World War II "golden era of flight" without any sense that she may have been something of an anachronism (Whyte and Cooper 1991, 212).

If the many postwar female pilots were to succeed in the marketplace, mobilization, organization, and leadership were necessary. Unfortunately, it was only much later, in the seventies and eighties, that Whyte began consciously to serve as a role model for female flight students and to think of others' professional ambitions. The idea of writing, perhaps inspirationally, came only very late in life. Sadly, at the crucial moment of the forties, Whyte, like Harris and Hart, was looking out only for herself.

The Airwoman Who Stayed the Course: Jacqueline Cochran

In stark contrast to the other great airwomen, Jacqueline Cochran never missed a beat in the postwar years. Cochran continued an incredible flying career, continuing to set records, often in military aircraft, into the sixties.

Once World War II and its immediate aftermath were over, air racing was widely expected to return to its prewar state. The hugely popular transcontinental Bendix race was revitalized with great anticipation and

Jacqueline Cochran with her P-51 racing plane, 1947. Smithsonian Institution Photo No. 86-533.

enthusiasm. With many very fast warplanes on the surplus market, it promised to be another golden era of racing. What could be more exciting than having Lightnings, Bearcats, and Mosquitoes, legendary war machines of great destructive power, competing against one another? It would be like the 1930s, only better.

Cochran obtained a P-51 Mustang, one of the most successful fighters of the war. Sparing no expense, she had the machine modified for top racing performance. It was widely expected to be among the most competitive. She joined a long list of other hopefuls, most of whom were similarly equipped with P-51s.

Cochran's careful attention to hardware and her tremendous flying skill did not disappoint these expectations. She came in second in the 1946 Bendix, beating all her male rivals but winner Paul Mantz ("Mantz Roars 435 Miles an Hour," 1946). However, things were not the same. To be sure, powerful airplanes had gone very fast, and oldtimers like Cochran were very much in evidence. But the winning machines were not even close to being the fastest things around. The future of absolute aviation speed records belonged to the most spectacular aircraft to emerge from World War II: the jets. Old-style races had to be thought of with the qualifier, "for propeller driven aircraft."

The old warbirds were clearly in trouble, but they evaded forced retirement by changing venue. Closed-course pylon racing had a great excitement value that faster, less maneuverable jets could not match. In fact, World War II fighter airplanes dominate this form of competition today.

Jet aircraft, of course, were and are tremendously expensive. For all practical purposes, no private parties could afford to field jet racers, even if they could have wrested them from secrecy-minded governments. The military owned the jets. Officers, who were more akin to well-trained graduate engineers than the swashbuckling, hard-drinking romantics that their prewar counterparts had been, would set the records. The 1946 race was the beginning of the end for the Bendix.

None of this detracts from the magnificence of Cochran's accomplishment. Indeed, this race is perfect evidence of her unique stature. The fact that she won second place might suggest that women were maintaining their place in aviation; however, Cochran was the only woman in the race. To be sure, another ex-WASP, Nadine Ramsey, had intended to race in a P-38 but did not compete ("Enters Bendix Race," 1946). This was a far cry from the multiple female entries of the thirties.

In the 1947 Bendix (for propeller-driven aircraft), two women competed: Jane Page Hlavacek, who came in ninth, and Diana Cyrus, who did not finish. Hlavacek's P-38 time was not at all impressive and she would have finished out of the money except for a special prize for the fastest woman's time ("Entries and Finishers," 1947). One wonders whether Cochran had by this time seen the handwriting on the wall for propeller-driven airplanes, for she did not compete in the 1947 Bendix. Instead, she loaned her finely tuned racer to a male pilot who made a fourth-place finish — at a considerably slower time than Cochran's in 1946.

Cochran was not quite ready to give up on the P-51, however. She used it to establish a number of speed records for propeller-driven airplanes, the most impressive for a closed-course run at 438 miles per hour in 1949 ("438 m.p.h. Mark for Propeller Plane," 1949). Like most of the records that Cochran coveted, this one was unqualified, worldwide for male and female pilots. Almost incidentally, her various P-51 trials gained for her the international absolute speed record for women, regardless of category.

By virtue of these feats, Cochran received considerable recognition from sources as diverse as President Truman and the Campfire Girls. She received the Harmon Trophy for "top Airwoman" in the United States, the Lady Drummond Hay Memorial Trophy from the Women's International Aeronautic Association, the French Aeronautical Medal, and the Hawk Award from the American Legion, to name a few. No female pilot in the United States even approached Cochran's accomplishments.

In May 1951, however, Jacqueline Auriol of France took the absolute women's speed record away from Cochran. Auriol was something of an upstart newcomer in that she had only recently learned to fly. She certainly did not have the veteran status of Cochran, but she had an immense advantage: access to military hardware. Auriol was certainly a skilled pilot — she was, in fact, a professional test pilot for the French government and went through all of the rigors of that occupation with flying colors. She earned her spurs and consistently performed at the level of her male counterparts. But she was, in addition, the daughter-in-law of the President of the Republic.

Auriol's combination of skill and clout led to a set of trials in a British-manufactured Vampire fighter plane. She was clocked at 509 miles per hour. Although Cochran often seemed to eschew gender records, she reacted strongly to Auriol's feat. She insisted that her own record was still intact, for she had flown a propeller-driven plane while Auriol had flown a new jet ("World Speed Mark for Women," 1951; "Says Her Speed Mark Was Not Broken," 1951).

Cochran did not spend much time in such pointless argument. Although she may not have had quite such intimate relationships with the politically powerful, she was certainly not without resources. She began to hound her contacts in the military for a powerful jet to reclaim the record for the United States (Cochran and Brinley 1987, 133). Through assertiveness, the help of well-connected financier husband

Floyd Odlum, and friendships with various Air Force and general aviation notables, Cochran planned an assault on the sound barrier.

She was trained by Chuck Yeager, the first person to beat the speed of sound. A Canadian-built F-86 Sabre with a powerful British-built engine was obtained on loan from the manufacturer. Raw ability triumphed again as Cochran dived the jet to the speed of sound on May 18, 1953, the first woman to do so. In the days that followed Cochran established several classes of speed records — unqualified and without respect to gender — with the same plane: 100-kilometer closed course, 500-kilometer closed course, and 15-kilometer straightaway. In the process, she recaptured at 653 miles per hour the international women's speed record from Auriol — who had upped her own record, in the meantime, to 540 ("Cochran Becomes First Woman to Break Sound Barrier," 1953; "Cochran Sets 2 More Speed Marks," 1953; Cochran and Brinley 1987, 280–281).

These accomplishments are especially impressive because Cochran was nearing fifty years of age; she was by no means the youngest of the airwomen. Auriol, meanwhile, was planning her own supersonic flight. Flying the new French Mystere fighter, she too broke the sound barrier on August 29, just three months after Cochran — and recaptured the international women's speed record at 688 miles per hour in the process ("Auriol Breaks Sound Barrier at 687.5 m.p.h.," 1953). This set off squabbles in the press about which records had been broken and which were still intact. This exchange, reminiscent of the parade of women's endurance flight records of the 1930s, continued well into the 1960s with each woman, able to dip into her respective country's military arsenal for appropriate vehicles, alternately upping the highest speed at which women had ever flown.

In a word, Cochran's postwar achievements are spectacular. No observer could deny that women could be successful in the new, fast-paced jet era of aviation. Cochran's record in the years after the war equalled or surpassed her accomplishments in the thirties. Without doubt her post-1944 accomplishments are more impressive than all but a few of the airwomen in their heyday. Given all that she accomplished during this period, the potential for leadership was great indeed.

Analyzing this record from the standpoint of its implications for the advancement of women, however, qualifies her achievement somewhat. The postwar story of Cochran, though it is extremely impressive, was not really an advance over what happened in the twenties and thirties; it basically repeated the earlier pattern with new and more technologically sophisticated equipment. Her work in record setting, like that of Whyte in closed-course racing, was a throwback to the demonstration stage. While it was obvious that demonstration of female competence had to continue, something more than simply broadcasting this message to the land was necessary. Either Cochran did not grasp this or she was not ingenious enough to cope with the demands.

Most of the sensitive airwomen — even Harriet Quimby, as early as 1912 — seemed to sense that widespread female professional success would one day occur. The WASP experience seemed to show that it was on the horizon by 1944. What might the most eminent airwoman, a person of great business, organizational, and piloting skills, a well-connected Washington insider, the past commander of the WASPs, have done? What would launch the professional stage? Several possibilities emerge from the experiences of other groups. Formal organization seems to be the key. Well-established groups can either provide education and training or direct their constituents to it. They can also provide valuable services, such as networking and nationwide job listings. Perhaps most important, they can shape public opinion and government policy by conducting aggressive public relations campaigns, lobbying public officials, and mobilizing their members as voters. All of these things in some way imply the formation of a community. Cochran, however, was strictly involved in solo work. Unfortunately, solo performances made their best contributions during the demonstration stage. This stage, despite Cochran's great performances, had waned.

Cochran was, without doubt, the premier female aviator of the day, but there were others. Auriol, her international rival, might have been a comrade. The T-6 racers were noisily if briefly present and Cochran herself had been instrumental in establishing AWTAR. Could she have traded upon the great respect that she had earned to promote and lead a community of female aviators? Could she have organized demonstrations of women's ability to fly instead of contenting herself with singular virtuoso feats? Could she have gone further and used her great popularity to begin and promote the professional stage?

Cochran did not do these things. Though she was photographed with Jacqueline Auriol and though the two were reported to be friends, Cochran never claimed fellowship. Auriol is scarcely mentioned in Cochran's writings (Cochran 1954, 245; Cochran and Brinley 1987, 281); Lettice Curtis, a British test pilot and former ATA girl who briefly held the women's world record in 1947 ("Government Test Pilot Makes New Women's Record," 1947), never seems to have penetrated Cochran's awareness. Given Cochran's indifference toward gender-specific events, it is not surprising that she ignored people like Harris and Whyte. Her involvement with AWTAR and the Ninety-Nines was not as an equal participant but as a remote Lady Bountiful (Douglas 1991, 70).

In terms of ability and performance, Cochran was well out in front of the postwar pack. Her many awards and recognitions demonstrate that she was regarded as exceptional by the lay community. Her thoughts and advice were often solicited. These are commonly regarded as evidence of leadership. Cochran, however, was not interested in empowering any followers. Like Whyte, she was in the postwar aviation milieu strictly for herself and was thus irrelevant to many of the needs of women in aviation. We will return to this theme in Chapter 5.

The Postwar Media

While it is clear that female pilots did not advance to a professional stage following World War II, we must still ask how they were viewed by the broader culture. Public attention and editorial support, although perhaps able only to extend the demonstration stage, would nonetheless have been extremely valuable. They might have prevented the erosion of gains already made until conditions more favorable for professional advance developed. Many factors suggest that press coverage was likely to be extensive. There was a much larger number of female pilots than ever before, trained in the WASP and other defense programs. Competitions and airshows for women were organized and venues like the Bendix were still available to women. The Cochran-Auriol connection promised an exciting combination of an international competition between colorful personalities and the growth of jet aircraft technology.

In point of fact, the information flow about postwar female pilots differed greatly from that which surrounded the airwomen. First, in the decade following the war female pilots rarely set their thoughts in print for general audiences. Only two books, appeared during this period: Cochran's *The Stars at Noon* (1954) and Hart's *I Fly as I Please* (1953). The former title, like Harris's *West to the Sunrise*, refers to an anomaly apparent only in special forms of airplane flight: in setting her jet records, Cochran had climbed so high that the atmospheric diffusion of light was attenuated even during the daytime — hence the visibility of stars at noon. Although Cochran's effort did receive a generally favorable review in the *Saturday Review* (Johnson 1954, 18), neither book was a literary masterpiece on the order of the works of Anne Morrow Lindbergh or Beryl Markham.

Others of this period, such as Harris and Whyte, would eventually write autobiographies. Dot Lemon, another T-6 racer, would some years in the future produce a fantasy narrated by an airplane rather than an aviator (1963). Molly Bernheim would tell the story of one of the few who was able to jump on the immediate post–World War II aviation bandwagon and live the life that optimists expected would be commonplace (1959). These, however, were all in the future. There is very little evidence that female pilots wrote for magazines and journals. Certainly there were no systematic attempts to attract a general readership, as had been the case with Amelia Earhart in *Cosmopolitan* magazine. Nor did postwar female pilots receive much exposure at the hands of other authors. Books about them were even more scarce than books written by them.

The most telling data, however, concern newspaper coverage of female pilots. An examination of the stories in the *New York Times* devoted to female aviators is instructive. For the decade following 1944, the average annual number is thirteen and in no year is it higher than twenty-six. Moreover, the count includes all stories about female aviators, with regard to any kind of activity. For example, stories of Cochran's financial transactions are included, as is a series about the rescue of Ruth Nichols, no longer an active pilot, at sea. Low though the total is, it thus overstates the attention paid to postwar female pilots' aerial exploits. As one might expect, Cochran is the most frequent subject of these stories, with Auriol as a strong second.

More discouraging, perhaps, is the lack of journalistic commentary. Only one editorial on female pilots appeared in the decade following World War II: a favorable one on Cochran shortly after she was awarded the Harmon Trophy in 1951 (Niekamp 1980, 144). Magazines and journals devoted slightly more attention to female pilots. Betty Skelton, a successful aerobatic competitor, was occasionally referenced in aviation publications such as *Flying* (Fuller 1949, 36ff.). The few professional women's journals did publicize AWTAR pilots as examples worthy of emulation (R. White 1949, 326–327), but an examination of popular magazines of wide circulation reveals little. *American Magazine*'s "Interesting People" column briefly noted Cochran (April 1947) and then Noyes (July 1948). *Life* published no articles at all on female pilots, but did display a photograph of Cochran (June 6, 1949) and another of a woman during her first solo flight (March 29, 1948). The *Saturday Evening Post* paid no attention, while *Colliers* emerged as champion with three articles, one featuring an interview with a very early airwoman, Blanche Scott ("Wing Talk: Lady Bird," 1947).

Novels for young people in the postwar decade continued to feature aviation, but female pilots were not among the cast of characters. On the other hand, the airline stewardess novel, a genre already discussed, proved to be alive and well. Helen Wells, author of a successful series about an adventurous and heroic nurse named Cherry Ames, presented the case for another supposedly gender-appropriate career in *Silver Wings for Vicki* (1947).

Interestingly, two nonfiction children's books on female pilots emerged shortly after the war. Each seems to be something of an anomaly, which may account for the fact that little else along these lines was to be published for many years. *Women in Aviation* was nominally very advanced in that it contemplated professional female pilots. It offered career guidance in light of the great new opportunities that the author thought had been created by World War II (Peckham 1945). She seems to have engaged in the wartime optimism about the coming peace, but failed to check the actual job market before rushing into print. *New Wings for Women* is a series of short biographies that are presented as object lessons for girls. This book may have been ahead of its time in that its purpose was to "chisel away at the wall of prejudice, block by block" (Knapp 1946). However, it was also behind the times in that the object lessons were the airwomen of old and not mid-forties contemporaries.

In short, the literature that surrounded the postwar female pilots was sparse, especially as compared with the late twenties and thirties. Nor was it terribly favorable, even when authored by female pilots themselves. To be sure, there are no examples of overt hostility, but explicit editorial support is almost completely absent. Finally, some of the material, though concerning female pilots, seems a bit beside the point; it focuses on the wrong people, does not have a clear message, or fails to square with reality. The mass literature of the time could not, of course, be expected to promote the development of the professional stage for female pilots, but it did not even do a very good job of maintaining the demonstration stage.

Conclusion

An era of fast-paced social change culminating in a massive, tradition-disrupting war; a long history of increasing women's self-consciousness, a well-known and highly praised group of very able airwomen: these defined the promise of the decade from 1944 to 1954. Late in World War II, editors and commentators confidently predicted that many female aviators, having clearly proved their worth, would join men as airborne exponents of the new, peaceful age. This did not come to pass.

Perhaps the original airwomen sensed that World War II was a watershed, that a new stage in women's involvement in aviation was necessary. Perhaps they knew that demonstrating flying prowess and hoping that others would be inspired was no longer enough. Perhaps they believed that their historic moment had passed. Even if they did believe these things, their complete withdrawal seems strange. Why did the airwomen not feel that they could change their spots and assume key roles in the next stage? Such a next stage, if there was to be one, clearly would have a strong dimension of professional advancement. Many of the airwomen had at least budding professional concerns. Many were motivated by more than the simple desire to exhibit skills. Few, however, took up the challenge.

If the airwomen felt they could not take on leadership roles in the new and changed circumstances, they might in some way have groomed others and remained on the scene as advisers or respected consultants. Failing this, they might have used their fame in order to attract attention to female pilots, in order to generate a modest and low-key public relations program. That they did none of these things left a great potential resource unexploited.

Though there were many female pilots during this period, few achieved prominence. Those who did were not prepared to take up the baton handed off (or dropped) by the airwomen. Some were not conscious of the obstacles that historically had been placed in the paths of ambitious and competent women — to say nothing of ambitious and competent pilots. Even those who were aware seemed to have a remarkably self-centered view of their own accomplishments. They were not interested in being exemplars or spokeswomen; their individualism precluded leadership.

The one airwoman who did not withdraw, Jacqueline Cochran, continued a spectacular career begun in the thirties. Though the tradition of demonstrating women's flying abilities was thus continued, Cochran was, in a sense, stuck in an obsolescent mode. If she was conscious of the professional aspirations of other female pilots, she gave no evidence of it. She did not grasp the importance of the next stage. The broader culture might have regarded her as a leader, but she failed to promote the achievement of other female pilots.

Even if the proper cast of characters had been present, the prospects for female pilots still faced great obstacles. Media attention was severely limited. Books, whether authored by female pilots or not, were infrequent, delayed, or only obliquely addressed to the issue of women in aviation. Newspaper coverage was thin and almost totally devoid of editorial support. In a word, following the departure of the airwomen, the cause of women in aviation completely collapsed.

❧ 4 ❧

The Long Term:
Promise Forgotten

In the short run, the promise of the airwomen clearly failed. Their conquest of the air, a brilliant display of women's abilities, somehow led only to stagnation after 1944. A demonstration stage, in which women showed the world tremendous aviation abilities, had been brilliantly executed but had run its logical course; a new, professional stage awaited development. Although the airwomen of the twenties and thirties were variously motivated, we saw that there were always some with clear professional aspirations. By the forties, a large number felt that flying could be an extremely satisfying job. A new age, it seemed, was dawning.

Though the call for professional advance was clear at least from the time of the Richey affair, very little of a concrete nature actually happened. Few airwomen seemed to understand the need for organization, strategy, and leadership; or if they did, they did not know how to respond. Established airwomen, even those who had spoken of professional ambition, failed to pick up on this important theme; most had withdrawn from public life and those who remained reverted to a brand of self-centered individualism more appropriate to the demonstration stage. Even the most significant organization of female pilots, the Ninety-Nines, failed to understand the emerging needs of its natural constituency; instead of preparing for concerted economic and political action, it rushed headlong toward the re-creation of a prewar past. Nor did a cadre of new leaders, unencumbered by commitments to the past, emerge.

Fortunately, the pattern of the late forties and early fifties also had limited historical staying power. Ultimately, the era of women's aviation did dawn; professional aspirations, decades late to be sure, have at least in part been vindicated. Today there are significant numbers of professional female pilots, some of whom work for major airlines. Professional female pilots' organizations have developed as well. There are even organizations of women who use flying as a tool to pursue other professional or business activities.

Similar developments have occurred in the armed services. Though explicit combat roles are still prohibited, women once again handle the most sophisticated military aircraft. Despite some male naval aviators who apparently think that their sexual organs confer on them the unique ability to fly from aircraft carriers, women now routinely take off from and land on the decks of ships.[1]

In some sense the logical implication of the airwomen's work is with us. Is it the case, then, that they had an important impact, albeit delayed? This is an eminently reasonable question, for great people are sometimes not recognized in their own time; only later is the value of their contribution understood. If this is the case, how has their influence been carried over the years?

There are at least three ways in which traditions established by the airwomen could have been kept alive. First, aviation historians, if they

---

[1] A recent magazine article on carrier operations explains the male naval aviator's cry, "Flare to land, squat to pee" (Moll 1992, 77). Its meaning comes from the fact that in a conventional landing, one establishes an angle of descent, sometimes rather steep, toward the runway. Near the ground the pilot attenuates this angle, greatly reducing the rate of descent and thus effecting a smooth touchdown. This action, which involves bringing the airplane into a slight nose-up attitude, is called "flaring." In aircraft carrier operations, a landing must be accomplished in very little space, so the luxury of smooth touchdowns is sacrificed. For all practical purposes, such landings involve no flare at all, just a fairly violent, steep arrival on the deck. The pilots who live by the slogan above are very proud of their mastery of this special technique and are correspondingly disdainful of those whose experience is limited to normal landings.

It is obvious that vulgar sexism in the aviation community is not dead. However, one can hope that this example represents a last gasp rather than a virulent resurgence. *Flying* magazine, which published the story in which the slogan appeared, did so as a matter of reporting on the current situation without conveying any approval of such attitudes. In fact, this magazine had recently featured a cover story entitled "Women with Navy Wings" (Laboda 1990), a favorable assessment of women's success in aircraft carrier operations.

were sufficiently sensitive, may have chronicled the airwomen. In this way, their stories would be established in an accepted literature and when at last the world was ready for them, they could be accessed. Second, this very same function could have been performed by agents other than specialists in aviation information. Writers interested in the status of women are no longer regarded as odd. Books of women's achievements and lists of notable women have appeared with some regularity in the last two decades. Indeed, increasingly well-organized disciplines, "Women's Studies" and "Women's History," have developed. It is certainly reasonable to think that workers in these areas would have wanted to preserve the memory of the airwomen regardless of the thinking of aviation historians. Both of these processes presume not only that writers have archived the traditions of the airwomen, but that those who came later, at least the contemporary female pilots, somehow accessed those records and allowed them to influence their lives.

Finally, it is possible that today's successful female pilots have been inspired by the airwomen through some process more subtle than literary or historical chronicling. For example, one can envision a sort of interpersonal communication across the generations. To be sure, there is a problem generation, that between the airwomen and the contemporaries. This is the group that opted out and from which, in a formal sense, we have heard little. Nonetheless, as anthropologists tell us, oral traditions can be transmitted across many generations. Perhaps the equivalent of stories told around a tribal campfire have preserved, for a crucial few, the exciting tales of the airwomen. As a variation on this theme, contemporaries may have made a special effort to locate predecessors on the path that they were motivated to take; a modest amount of probing into the obscure would have uncovered the airwomen.

Aviation History and the Airwomen

Dominick A. Pisano and Cathleen S. Lewis (1988), having prepared the most extensive annotated bibliography on the history of flight, note that historical interest in aviation has only recently begun to develop. While there have been many attempts to provide descriptions of given events or glimpses of particular personalities, few contemplative, interpretive, or theoretical works have yet been written. In general, writers have not yet attempted to integrate the various strands of inquiry, to discuss long-term implications, or to view aviation as an important technological and social phenomenon with continuing meaning for our culture.

Aviation history, like most new fields, is thus somewhat chaotic. Much of it consists of small bits of information presented for their own sake rather than as a contribution to an orderly body of knowledge. Isolated facts, specialized information, or exotica are the order of the day. Intriguing stories of the past, tales with inherent, usually "human interest" appeal, make up the bulk of the literature.

Consequently, Pisano and Lewis are able to list very few general aviation histories among their thousands of entries. This rather sparse picture is confirmed by the catalog of one of the great aviation libraries in the world, the National Air and Space Museum in Washington. Titles having to do with the general history of aviation are shown to be in short supply.

Though the list may be short, it spans quite a range, from formal studies by academic historians through journalistic reports to picture books. While it is probably premature to identify conventions or canons in this writing, it is possible to examine the emerging overall picture of aviation that they present. And, of course, it is of singular interest to consider the degree to which female pilots are regarded as contributing to that picture.

Roger E. Bilstein's *Flight in America 1900–1983* (1984) is a true scholarly work written by a professional historian. This book not only surveys "principal technological trends" but also investigates the "social, economic and political aspects" of aviation. It asks where we have been and where a continuing technological tradition might take us as a culture. Indeed, it is just the sort of book in which female pilots need to be discussed if they are to be viewed as socially important.

Bilstein mentions six airwomen by name: Quimby, Law, Stinson, Earhart, Nichols, and Cochran. In so doing he acknowledges women as among the earliest pioneers, as involved in postbarnstorming economic development, and as members of the WASPs (22, 76, 84, 163). There is a photo of Earhart and several other airwomen at the 1929 National Air

Races, as well as mention of the WASPs as ferry pilots in the caption of a photo of a World War II fighter plane. Moreover, Bilstein notes a 1916 discussion in the *New York Times* on appropriate aviation terminology. This newspaper had explained to its readers that not all pilots were men, which raised the question of whether gender-specific terms like "aviatrix" were appropriate. Bilstein reports with apparent approval that the paper eventually settled on the term "aviator," which it regarded as gender-neutral.

While Bilstein takes care to include women, the specific entries are brief and the intent is sometimes not clear. For example, Bilstein reports that Nichols had a transport pilot rating. This is perfectly true, but Nichols was not particularly well known for her transport work. The WASPs are praised for competent discharge of their many duties, but Bilstein claims that they did transatlantic ferrying throughout World War II, which in fact they did not.

Bilstein spends several paragraphs discussing the early introduction of stewardessing on the airlines, and he describes a group of 1930s stewardesses appearing in a photo as "alluring." While his consideration of the economics of early commercial aviation certainly demanded discussion of stewardesses, it is unfortunate that the development of this role for women received more attention than all of the references to airwomen combined. In short, this professional historian has acknowledged the airwomen in a direct and positive way. However, that acknowledgment seems a little perfunctory and perhaps just a bit strained.

Other serious histories of aviation have been even less generous to the airwomen. *Two Hundred Years of Flight in America: A Bicentennial Survey* (Emme 1977) consists of nine essays on such topics as military aviation and commercial aviation. The thrust is clearly much more technological than that of Bilstein and, though explicitly economic, less social or political. None of the nine contributing authors mentions the airwomen at all. Oliver Stewart's *Aviation: The Creative Ideas* (1966) deals with various aviation accomplishments, especially technological ones dealing with power plants. No airwomen cross the pages of this book.

Lloyd Morris and Kendall Smith's *Ceiling Unlimited: The Story of American Aviation from Kitty Hawk to Supersonics* (1953) was apparently written for lay audiences. Its general organization is similar to that of

Bilstein; it recognizes several stages of aviation development not unlike those spelled out in Chapters 1 and 2 of this book. These authors miss some good opportunities to discuss airwomen, excluding them altogether from a section on barnstorming. Only two airwomen are mentioned by name, Earhart and Cochran; the entry on Earhart is rather standard, while a paragraph is devoted to Cochran's various records. Although praise is implied by the very fact of their inclusion, the authors do not suggest that they were major players of any kind. Similarly, the eleven lines about the WASPs that grace the chapter on World War II do not suggest much of a contribution. If Bilstein's treatment was brief, Morris and Smith's was abrupt.

Hendrik deLeeuw's *From Flying Horse to Man in the Moon* (1963) is a poorly integrated general history. It contains fifty-eight discrete, topical chapters in less than three hundred pages. There is an uninsightful discussion of Earhart and a two-page chapter devoted to "Ladies of the Air." While several female pilots are referenced, the entries are often inaccurate and no interpretation is offered.

As one moves into more informal, "journalistic" history, an enterprise likely to depend more on individual personalities and the use of photos, acknowledgment of airwomen does not improve. Perhaps the attention that female pilots are likely to generate on account of their rarity is tempered by the inherent conservatism of many such publications. For example, coffee-table books often detail exotica, but cannot diverge much from the sales-driven cultural mainstream. Indeed, one could argue that such books are an excellent barometer of what has arrived in the cultural mainstream.

The American Heritage History of Flight (Gordon 1962) shows this very clearly. *American Heritage,* essentially an establishment-oriented magazine, has long presented an idealized vision of the American past. This large-format book was no doubt intended to be entertaining and to make its readers feel good about being American. It probably succeeds on both counts. In many ways, it is a good compendium. Its section on "exhibition fliers" details many of the earliest pilots and their clear performance orientation. The Moisant group is mentioned, as are the Stinsons, but their inclusion seems to be dictated by a family connection. Mathilde Moisant accomplished much in her own right, but she and her brothers are, with some accuracy, described as a team. Similarly, the Stinson sisters were part of a business operation that featured their mother and their brother as well. *American Heritage* approves of family ventures; via this mechanism some early airwomen may have achieved a degree of immortality. In addition, this book briefly mentions Ruth Law's 1916 record distance flight and it features a photo of her "racing" against an auto. A photo of Harriet Quimby also appears, but without any comment in the text.

Chapters on barnstorming and developments of the twenties avoid mention of the airwomen except for a photo of actress-turned-pilot Ruth Elder. The section on the thirties is very diverse, focusing on technological development, record setting, and preparation for war. Earhart receives some attention here as does, very briefly, Ruth Nichols, as the holder of altitude records. Not surprisingly, the emergence of stewardesses on the airliners is also reported as an important development. Finally, in the discussion of World War II there is a brief quote by Jacqueline Cochran.

A volume that depends entirely on photos is John Taylor's *Flight: A Pictorial History* (1974). This item, which has a worldwide rather than a strictly American focus, contains 652 plates. Included are photographs of Earhart, Jean Batten of Australia, and Amy Johnson of England, all well-known female pilots of the airwomen's era.

The most extensive popular history of aviation is the Time-Life series, Epic of Flight. Prepared between 1980 and 1983, this effort grew to twenty-two volumes, each about two hundred pages long. While the organizational principles that led to this particular list of titles are not always clear, the scope is impressive. Groups of people, particular time periods, and types of aircraft are the bases for the volumes, which extend beyond the American experience. Given the nature of Time-Life, Incorporated, it should be no surprise that these books depend heavily on photos and illustrations of the type one would find in high-circulation magazines. Nonetheless, these books are well-researched and do a good job of presenting basic factual information, even though there is much emphasis on unusual detail, adventures, and the exploits of colorful personalities.

How did Time-Life, a decade ago, view the airwomen? First, many of the titles could not be expected to discuss the airwomen at all. One on early ballooning covered a period prior to the emergence of the airwomen, as did one dealing in the technological effort that culminated in the Wright brothers' success in 1903. Others, dealing with events in Britain, Germany, or the Soviet Union, would by definition exclude the airwomen. Still others treat of topical areas, such as dirigibles, helicopters, or aircraft carriers, the development of which did not involve airwomen. Nonetheless, extrapolating from titles alone, at least a third of the volumes could reasonably have considered the airwomen.

The First Aviators (Pendergast 1981) covers the history of heavier-than-air flight until about 1914. Its focus is on showmanship, competitions, and the technological developments that anticipated the many aeronautical advances of World War I. Dashing and heroic pilots are everywhere in these pages, including the Moisant International Aviators. The featured actors in this group are, however, the Moisant brothers. Recall that this team had three women: Mathilde Moisant, Harriet Quimby, and as a lesser player, Bernetta Miller. This book omits their involvement, however, focusing instead on the brothers' exciting history, including a stint as entrepreneurs and soldiers of fortune in El Salvador. The Moisant brothers deserve attention as representatives of a distinct age in the development of aviation, but it is regrettable that their female colleagues did not pique the curiosity of the Time-Life editors as they sought to describe this period.

The First Aviators mentions Harriet Quimby in a short sidebar about developing safety features, but she appears only as an anonymous victim. The Baroness de Laroche, the first woman in the world to earn a pilot's license, is incidentally mentioned in a caption to an illustration; an early woman passenger is briefly mentioned in the text. An unidentified aviatrix appears in a reproduced poster of that era. Other than this, the first aviators, by this account, are exclusively male.

More women are referenced in *Barnstormers and Speed Kings* (P. O'Neill 1981), but most of these are not pilots but performers. There are photographs of, among others, Gladys Roy, the aerial Charleston dancer. O'Neill does credit Mabel Cody as a major operator of 1920s airshows, recognizing business acumen if not flying ability. The pronounced neglect of women in this book is perhaps most clearly shown, ironically, at the one point where really great achievements are noted. There is a complete list of winners of the prewar Bendix transcontinental races, accompanied by photos of each pilot. True to historical fact, Thaden and Cochran are listed and illustrated, but there is no mention of them in the text whatsoever. Women of course did not win as many races as men in the twenties and thirties, but O'Neill was certainly persistent in practicing the art of understatement.

David Nevin (1980) prepared a volume about transatlantic, round-the-world, and other early long-distance flights. This book, called *The Pathfinders*, covers roughly the period from the end of World War I to the mid-thirties; it starts with the Atlantic crossing of two U.S. Navy sea-planes in 1919 and ends with Wiley Post's second circumnavigation of the globe in 1933. The great majority of the people making such flights were indeed men, so one would not expect women to be prominently featured. Nonetheless, the degree to which women are ignored in these pages is startling. The name of Vera Mae Dunlap, a wing walker in a barnstorming troupe which once included Lindbergh, appears in a reproduction of an early twenties handbill. Just as Pendergast found a female victim in Quimby, so Nevin tells us of Princess Anne Lowenstein-Wertheim, a small-time pretender to royal lineage. The Princess arranged a westbound transatlantic flight in 1927, desiring to be the first woman to cross the Atlantic by air. She and her two hired pilots set out in some secrecy, but failed to arrive in North America and were never heard from again. A single sentence acknowledges Earhart's 1928 passenger crossing and the records of Lady Mary Heath and Lady Mary Bailey. The latter two made spectacular solo flights between their native England and South Africa in the twenties. Women are made to seem very small players indeed.

The Explorers (Jackson 1983) discusses various adventurers who, beginning in the twenties, used airplanes instead of the traditional canoes, dogsleds, camel caravans, safaris, or ships to visit remote corners of the world. This author discusses three American couples who uncovered exotica by air. Richard and Mary Light did spectacular aerial photography in Africa in the thirties; their 1937 photo of previously mysterious Mount Stanley is a classic. Martin and Osa Johnson were very traditional African safari types who, with a large crew, produced books and films of their adventures. They flew in an amphibian painted with giraffe-like mark-ings. The Lindberghs' well-known 1931 flight to Japan is also noted. However, the women in these dramas were passengers and secondaries.

It is unfortunate that Laura Ingalls's solo flights around South America, for example, were not thought to fit the mold of *The Explorers*.

Occasionally, this series does bring to light female pilots who are not widely known. This is the case in *Bush Pilots* (Editors of Time-Life Books 1983), a collection of short pieces about various pilots who probed the remote and inhospitable areas of the world. These rough-and-ready types are perhaps best known for servicing remote outposts in Alaska and the Yukon, or connecting communities in sparsely settled central Australia with civilization. Few airwomen, aside from Marvel Crosson, were bush pilots. Crosson is not mentioned, but Elizabeth Greene is. The latter did her bush flying after World War II in isolated areas of the Yucatan. With an old Waco biplane, she supplied a group of missionaries. Greene qualifies as an airwoman because of her service with the WASPs.

More often, books of this series do not mention the airwomen at all. Donald Dale Jackson's (1982) discussion of early airmail pilots mentions neither Stinson nor Law, legitimate pioneers in this area. Edward Jablonski's *America in the Air War* (1982) summarizes national involvement in World War II, but avoids mentioning the WASPs at all. Indeed, the only appearance of a woman is a photo of a production worker in an aircraft factory.

In some cases the series, perhaps unwittingly, presents stereotyped or even negative images of women. Oliver E. Allen's *Airline Builders* (1981) would not be expected to feature airwomen because few of them took key roles in that development. This book does, however, discuss women as stewardesses and passengers, by implication excluding them from a more active role. Richard P. Hallion (1983) wrote about the history of designers and test pilots, and not surprisingly, mentions no women; although there were female test pilots, few achieved prominence in that role. The only reference to an airwoman, a photo of Pancho Barnes, does not in any way acknowledge her career as a pilot. Instead, she is shown as the proprietor of an establishment called the Happy Bottom Riding Club. After she retired from flying, Barnes established this local hangout near Edwards Air Force Base. Since that was the location of many postwar jet aircraft tests, many well-known male test pilots like Chuck Yeager patronized the place.

Sterling Seagrave (1981) put together a series of tales about pilots who had served as mercenaries in various wars. Few, if any, airwomen undertook this role, and so they are absent from this volume. However, there is a lengthy story about a U.S. male who flew for the Republican forces in the Spanish Civil War. At one point, he was shot down and captured. A lady friend, a variously talented entertainer, generated great publicity in attempting to secure his release. Pleading letters to General Franco, which included photographs of her in scanty attire, were apparently a key element in her strategy. Seagrave goes into some detail, reproducing one of the photographs. Apparently the poor pilot, once released, found his lady friend had developed new interests. While all of this was no doubt true, it is a curious inclusion in a history of aviation. Certainly, it suggests that women's contributions to that history are inherently quite different from those of men.

As if in compensation for the limited treatment of the airwomen in these volumes, the series includes a title, *Women Aloft* (Moolman 1981). This excellent book covers many of the airwomen, particularly the earliest and those who strove after various speed, altitude, and endurance records. A long section on Earhart, though well-written and accurate, conveys nothing new. Valerie Moolman's discussion of the WASPs is enhanced by comparisons with the British ATA and the Soviet experience with female combat pilots in World War II. The latter are made especially interesting by virtue of information from primary sources. While one could quibble with Moolman over what she chose to include or exclude, her work accurately portrays the airwomen and fairly recognizes their accomplishments. Though it is not brilliantly interpretive, this work does effectively argue that women should have a secure place in aviation history.

In a sense, *Women Aloft* redeems the series and compensates for neglect in the other books. At the same time, however, this volume symbolizes that the airwomen have not been integrated into "regular" aviation history; they require a special volume because they seem to fit awkwardly or not at all into the other topically organized works. Does separate consideration imply inferiority, or is it a well-deserved distinction that is likely to bring appreciation and acceptance? Earlier chapters discussed the implications of separate record categories and races for women. Just as the merit of these moves was debatable, so is the appearance of a separate book in this series. The airwomen penetrated the historical consciousness of the Time-Life editors, but apparently did not seem worthy of consideration in the same venue as their male counterparts.

Historians, then, have had a rather mixed reaction to the airwomen. Some have ignored them entirely, while others have paid them some tribute and acknowledged their contribution. Often, however, that acknowledgment seems brief and grudging. Moreover, the historians appear to follow popular culture in concentrating on Amelia Earhart. There is little recognition of the many dimensions of the airwomen's achievements. Even more problematical, these historians repeat a feature of the mass media of the twenties and thirties: they often reinforce the notion of separate aviation roles for women, such as stewardessing and wing walking.

In a word, the airwomen have not captured the hearts and minds of the chroniclers of the past. In turn, it is difficult to see how the hearts and minds of readers of this aviation history could be captured. In a search for roots, ambitious women already inclined toward aviation might regard some of these materials as pay dirt, particularly the Moolman volume, but it is difficult to think of the total available product as the basis for broad inspiration.

Women's Studies, Women's History, and the Airwomen

Contemporary observers in women's studies and women's history would not be surprised to learn of the minimal attention aviation historians have given the airwomen. Their untold story would likely be viewed as just one example of a very broad problem. Indeed, one of the prime motivations for the development of such fields as women's studies and women's history has been a perceived lack of attention to women. It is a commonplace assertion that most writers and historians have been men. By imposing a peculiarly male interpretation on the events they see, the argument goes, they have systematically ignored or diminished women's participation in important events. A male world is described for us; we read male history. Regardless of how much they achieve, women have been barred from hero-hood. As a result, there are few female role models for girls.

As these scholars try to redress the balance, as they properly call attention to a more accurate and elaborate picture of women in our culture, will they also act as creators of an airwomen's legacy? There are two logical places in which to look for an answer.

The first is in what one might call biographical directories or biographical dictionaries. Presumptive lists of what is notable, or who is notable, are a long-standing tradition. Everything from serious scholarship to snobbish social registers have relied on some sort of exclusive published compendium, and various constituents have predictably worried about whether what, or who, was on or off the critical lists. Probably the best-known biographical directory is *Who's Who in America*. While the primary purpose of this publication may be to gratify the egos of those included, it nonetheless is a valuable reference source. Specialized *Who's Who* volumes reference particular populations, such as scientists or those living within specific geographic regions. There has been a *Who's Who of American Women* since 1958.

Who's Who and its ilk refer only to the most contemporary notables; there are periodic new issues and people are systematically added as they achieve status and dropped, usually following retirement or death. As might be expected from Chapter 2, the airwomen were well represented in biographical directories of their time. *American Women*, very similar to the modern *Who's Who*, was published annually from the mid-thirties through 1940. It was very general and tried to include "all the women who figure prominently in our national life today" (Howes 1981, vii). This publication lists women by, among other categories, occupation. The category of "pilot" lists twelve airwomen, ten prominent and two locally known. Three other women, officers of aviation organizations, are also listed. Women in aviation is a meaningful category in this volume.

Contemporary note is, however, not our main interest here; more to the point is the question of persistence over time. Historians in particular have had similar quests and have created biographical dictionaries of those whose importance transcends their own eras. Just as directories of contemporaries have proliferated through specialization, there is no single accepted listing of historical notables. It is only natural that those who feel that the dominant culture has excluded important people would prepare their own biographical dictionaries; indeed, there are a number that reference women exclusively. Perhaps the most scholarly is the four-volume work, *Notable American Women* (James, James, and Boyer 1971; Sicherman and Green 1980). Clearly motivated by the desire for historical completeness, this effort appeared under the auspices of Radcliffe College. It attempts to identify important women from the colonial period — the formal start of coverage is 1607 — to the date of publication. Indeed, a great many women in a great number of fields are identified, for there are nearly three thousand pages of brief entries. Nor does it slight the period in which the airwomen lived, including such accomplished women as Margaret Sanger, Grandma Moses, and Rachel Carson.

The effort that went into *Notable American Women* appears to have been both thorough and sensitive, for among contemporaries of the airwomen it identifies a great many entrepreneurs, a number of physicians, more than twenty painters, some eighteen historians, and even four astronomers. It seems as though it would be an excellent place in which to find records of early women aviators. A hint that this might not be so, however, appears when we note the classification scheme for the entries in these books; while there are specific categories for entrepreneurs, physicians, painters, historians, and even astronomers, there is no discrete category into which airwomen would fit. Only three airwomen, grouped generally, grace these pages. Predictably, there are several paragraphs on Amelia Earhart. Very brief entries on Ruth Nichols and Phoebe Omlie also appear; it is not at all evident why these two were chosen while others, most obviously people like Louise Thaden or Jacqueline Cochran, were not.

Some women's biographical dictionaries are meant as more than scholarly reference sources. Their compilers openly hope to supply grist for the mills of feminist social change. *The Continuum Dictionary of Women's Biography* (Uglow 1989) attempts not only to supply information otherwise unavailable, but, in thousands of short entries, to show "strength in action" and to answer the "request for heroines." Uglow considers about as many fields as *Notable American Women*, but her time frame is narrower (the past two centuries) and her geographic scope is wider (North America, Europe, and the British Commonwealth). As far as criteria are concerned, the volume includes women "whose role in history, or whose contribution to society or use of talent would be remarkable regardless of their sex" (vii).

An excellent classification system groups many subheadings under four main themes: public life, cultural life, physical achievements, and dynamic characters. In some six hundred pages a large number of Americans from the airwomen's era are noted — including, appropriately enough, everyone from Lizzie Borden to Elizabeth Arden to Frances Perkins. Joining these are vast numbers of entertainers, twenty-two who made careers in medicine or the sciences, eleven who succeeded in journalism or other communications fields, five photographers, and even three mountain climbers — all Americans active sometime during the 1912–1944 period. The "Aviation/Space" category includes a few recently active Americans such as Sally Ride and Jeanna Yeager, as well as a few non-American contemporaries of the airwomen; but out of all the American airwomen, only Earhart and Cochran appear. While aviation as a concept appears to have achieved some stature, the airwomen remain, except for a perfunctory bow, unnoticed.

The Women's Book of World Records and Achievements (L. O'Neill 1979) is similar in purpose; it intends to provide "heroes to set young women free." The volume, some eight hundred pages long, is marred by the fact that some of the entries read a bit like answers in trivia games. The classification scheme is solid, however, featuring such concepts as politics, community, science and technology, medicine, and law, with various subdivisions. Under "Sports: Airplane Pilots," a few contemporaries are noted, as is a single airwoman. Jacqueline Cochran receives one paragraph. To be sure, O'Neill has a group she calls "Far Out Women," apparently otherwise defiant of classification in her generally topical system. Here there is an entry on Earhart of less than a page, as well as one-liners on Tiny Broadwick and WASP Ann Carl. Broadwick was probably the first woman to make a parachute jump, but O'Neill's suggestion that she was the first person to do so is erroneous, as are the reported dates. Carl, who was assigned to a test unit, took a Bell P-59 aloft in 1944, becoming the first woman to pilot a jet plane. Under the military classification there is less than half a page devoted to the WASP organization.

World Records and Achievements devotes more attention to the air-women than either of the two biographical dictionaries discussed previously, but when one compares other categories of women contemporary to the airwomen, this degree of attention looks like neglect. Among news reporters of this period, such giants as Nellie Bly, Dorothy Thompson, and Pulitzer Prize winner Anne O'Hare McCormick are discussed. Along with these, however, are Winifred Bonfels, Adela Rogers St. John, Sigrid Schultz, Therese Bonney, Helen Kirkpatrick Milbank, Doris Fleeson, Elizabeth May Craig, Margaret Bourke-White, and Esther Van Wagoner Tufty. To be sure, in many people's minds, journalism is dramatic and exciting, so perhaps it is not unreasonable to include many practitioners of the craft. But even the category of astronomers lists five women of the 1912–1944 era, each rating their own substantive entries: Annie Jump Cannon, Antonia Maury, Margaret Harwood, Marjorie Meinkel, and Cecilia Payne-Gaposchkin.

Of course, none of this is meant to disparage the women noted above, nor to trivialize the fields in which they made their accomplishments. Indeed, these women deserve praise for success in traditionally male fields during an era when this was especially difficult. The important point is that many women whose names are not household words or who worked in very specialized fields readily attract more attention among women's dictionary biographers than do the airwomen.

The *Handbook of American Women's History* (Zophy 1990) is similar to the works cited above, but contains entries on key concepts and important events as well as notable individuals. Once again, the scope in terms of both time frame and substantive topics is immense. Angela Howard Zophy strives to be comprehensive. A half-page entry on aviation in general describes a few highlights of the airwomen's era. An entry of similar length on the WASP experience appears in the section on the military, while an entry on the Ninety-Nines fills half a column. In addition, specific airwomen are referenced: Earhart is the subject of a full-page story, while Cochran and Quimby rate a half-page each. Katherine Stinson is recognized in a shorter piece. In total, the equivalent of more than four full double-column pages are devoted to the airwomen and their work. While this is considerably less than 1 percent of the total content of the book, it is nonetheless far greater recognition than that offered by any other similar publication.

What has produced this satisfying result? This question cannot be answered in any definitive way, but it is clear that the editor interacted with observers outside the historical mainstream. Most of the entries on the airwomen are written by Claudia Oakes, recently on the curatorial staff of the National Air and Space Museum. Though she has written monographs published by the Smithsonian, Oakes is a museum information specialist, rather than an established scholar. Somehow, cross-fertilization has occurred and produced a rather different outcome.

A second and generally more important product of women's studies and women's history are narrative works. These range from detailed biographies of specific individuals to excursions into feminist social theory. There are obviously many publications that could not be expected to reference the airwomen in any event; by definition, works on colonial witchcraft, contraception, and educational trends, though highly relevant for women today, have few interfaces with aviation.

On the other hand, in most branches of historical study, there are works that offer comprehensive, integrated descriptions of where we, as a culture, have been, and where we might be going. They seek to identify causes, to tell us what important happenings really mean, to put various factual pieces together into a comprehensive picture. Of course, we are interested in those whose specific goal is to paint women into this picture. The airwomen could be expected to appear in at least some works of this type.

William L. O'Neill undertook to write a comprehensive history of feminism in America (1971). This approach is quite different from that required for a comprehensive history featuring prominent women; there is considerable emphasis on those women who were active feminists, who vigorously promoted social action and wrote cogent, well-reasoned theoretical tracts. This sort of feminism, perhaps to be spelled with a capital "F," implies much more initiative than, for example, serving as a role model or simply being successful. The late nineteenth and early twentieth centuries formed a period of intense feminism associated with Progressivism, Prohibition, international peace conferences, suffragism, and various reform movements born of the distressing conditions of the early industrial

United States. Many historians, among them O'Neill, believe that aside from achieving suffrage, the feminist movement stalled just about the time that the airwomen were coming to prominence. By comparison with the heady and exciting decades just before, they view the airwomen's time as one of stagnation and apathy. In this view, not much of merit happened during the twenties and thirties, and for that matter until the late sixties, when another round of feminist activism took hold. Given this, it is not surprising that O'Neill and others like him do not mention the airwomen. Not even Earhart, who was nominally a member of the National Women's Party, a bit of an organizational throwback to the radical activism of a past era, attracted his attention.

Barbara Harris's study of women and the professions in U.S. history (1978) seems to share a similar disappointment with the twenties and thirties. Harris indentifies several stages of women's professional lives over the last two centuries, calling the 1920–1945 period, "The Vote but Not the Answer." She does not regard women's professional accomplishments during these years as particularly meaningful; instead, she focuses attention on the lamentable decline in radicalism of oldtime reformers like Charlotte Perkins Gilman and Antoinette Brown Blackwell. Harris views this as an era of setbacks, epitomized by the conflict over and ultimate failure of the first Equal Rights Amendment. This radical proposal was, in many ways, similar to the Equal Rights Amendment of the 1980s. Its prospect terrified not only men, but also many women who feared the loss of special, protectionist legislation of the Progressive years. While Harris discusses law, medicine, the ministry, and business, she ignores aviation. Having developed in such a dismal period of history, this profession and its exponents, the airwomen, receive no attention at all.

William Henry Chafe (1972, 1991) also attempts to show how American women's roles have changed with time. His analysis of the period from 1920 to the present emphasizes general social, economic, and political factors, rather than feminist activism as such; but he, too, eschews discussion of particular personalities or possible role models, especially when discussing the twenties and thirties. Here the main focus is on powerful organizations, the economy, industry, and college preparation for careers. These matters are often addressed through the use of statistics and poll results. The discussion of World War II, which Chafe takes to be the real watershed event in changing opportunities for women, focuses on female production workers and their demand for child care facilities. Not surprisingly, given the main features of Chafe's argument, there is but one reference to an airwoman in this book: a passing and inaccurate comment on Jacqueline Cochran's "ferry service" during the war (Chafe 1991, 132).

Carol Ruth Berkin and Mary Beth Norton (1979) edited a volume with the simple title, *Women of America: A History*. Their purpose was to present examples of broad trends from colonial times to the twentieth century, but not to be comprehensive. Apparently, the book is intended to convey the idea that there is an identifiable women's history. They present papers on a variety of topics, though how those topics were selected remains murky. To be sure, themes that are by now familiar emerge: family and birth control, education, work, and reform movements. Again, various historical periods are discussed. There is no particular reason why a discussion of the airwomen would not have been appropriate in addressing their era, but they do not appear. Instead, this era is represented by an essay on a women's organization in the South, the Anti-Lynching Association, and by a paper on how women's images were manipulated for political purposes in World War II propaganda. In point of fact, the airwomen had simply not penetrated the consciousness of the many writers represented in this volume.

Sheila Rothman (1978), like others, envisioned an orderly progression of stages that could describe the status of American women over time. These stages, defining the period from 1870 to the present, derive from prevailing "ideals and practices," the consensus of the culture. Women moved, during the earliest years, from a condition in which "educated motherhood" would have described typical reality through several intermediate levels to today's stage of "woman as person." Perhaps it is because the airwomen were not typical of the stage in which they lived that Rothman gives them no attention at all. She titles their era the "Politics of Protection." Few airwomen could be regarded as exponents of this form of Progressivism; indeed, it may be that many airwomen had already grasped the notions underlying the "woman as person" stage to come.

The pattern in these books is repeated in many others not reviewed here. The central point, however, is well demonstrated by a comprehensive history of American women that does mention an airwoman, at least incidentally. Nancy Woloch wrote *Women and the American Experience* (1984) with the idea of bringing forth a new cast of characters on the historical stage. Since women have traditionally been in the background in historical studies, we have been deprived of potentially valuable insights. Woloch discusses Earhart, but in a very strange way: she is not represented as an accomplished pilot, but rather as a cult figure in some ways symbolic of the personal freedom inherent in the notion of flapperism. The most telling comment is that all of Earhart's accomplishments did not create any jobs for women (390). In Woloch's view, focused as it is on immediate conditions, Earhart and other airwomen are simply irrelevant. In this book, the important issues during the twenties and thirties were the economic and social ones springing, for example, from the Great Depression and Margaret Sanger's birth control initiatives.

All of these works, though clearly concerned about women and mostly written by women, focus on massive issues of overriding and immediate social or economic importance. They look for intellectual leaders who provide a theoretical basis for a coherent social movement; they admire activists and thinkers who, even when they fail, seem to be working in accord with some kind of meaningful revolutionary plan. To these authors, potential role models who were out flying airplanes seem to have been passive, uncommitted, unenlightened drones.

Of course, some writers have undertaken specifically to look at women's history during the twenties, thirties, and forties. Such authors obviously could not disregard the airwomen because of the purported insignificance of the period in which they lived. Moreover, many of these historians seem inclined to deemphasize the role of feminist activism or the predictions of economic theory in favor of everyday events in the lives of ordinary citizens. These events, they believe, need to be studied even if some observers would rather ignore the era in which they occurred. Dorothy M. Brown is critical of the work we have examined in the paragraphs above: "Historians of the women's movement carried the story through the Progressive Era and the suffrage victory and stopped" (1987, xi). While mainstream male writers may have ignored women altogether, historians specifically focusing on women may have committed a parallel sin; by neglecting postsuffrage women, they have consigned them to a lost generation. Brown and others intend to bring the lost generation to light.

Brown regards the twenties as an extremely important time for women. It is true that radical feminism went into eclipse, that dramatic political advances did not follow suffrage, that equality in the workplace was not achieved, that women were badly divided about social reforms. Nonetheless, women acquired college educations in numbers never before dreamed of. They entered the work force — albeit in "feminine" occupations — in droves. Above all, through a kind of personal achievement activism, they eroded a nineteenth-century traditional view of women's social roles that had, despite its anachronistic character in an urbanizing, industrializing nation, persisted well into the twentieth. In short, women may have been diverted, but they persisted and laid important groundwork. In Brown's view, the period produced a corps of exciting role models — like poet Edna St. Vincent Millay and attorney Mabel Walker Willebrandt — who set the course for future generations.

In this light, Brown does pay some attention to women's aviation in the twenties. She focuses her discussion on Earhart's 1928 trip across the Atlantic. Earhart is mentioned in the same breath as anthropologist Margaret Mead and athlete Gertrude Ederle, whose swim of the English Channel in 1926 was the first for a woman and the fastest on record regardless of gender. Brown argues that "Earhart's feat continued the erosion of the lingering true-woman stereotype" (44). While Brown's discussion is consistent with the meaning of the airwomen articulated in this book, it is nonetheless a bit disappointing. The entire entry is only two pages long, and others whose accomplishments were so much more dramatic than being a passive passenger are not mentioned at all.

Susan Ware's (1982) book on women in the thirties is similar in its basic thrust. That decade was dominated by the disaster of the Great Depression and the attendant hard times. Ware's title, *Holding Their Own*, reflects the economic concerns of all American women of this decade. Unlike many historians, however, Ware focuses explicitly on popular culture. Movies, radio, comics, and popular literature expanded tremendously during the thirties, due in part to new electronic and publishing technologies. This may have been extremely fortunate timing, for they provided a welcome diversion for a hard-pressed public. Ware argues that despite massive unemployment, hunger occasionally to the point of starvation, and deteriorating health, Americans never lost their composure during the Depression. They banded together in response to the crisis. There was never any serious attack on the basic social system; revolution was never a viable alterative, despite a fair number of assorted radical advocates. According to Ware, part of the reason for this benign public response was that people were able to derive real gratification from Charlie McCarthy, Katharine Hepburn, and Mickey Mouse. Mass entertainment, really a new innovation, made life tolerable. Because everyone could share it, popular culture was an important social cement. Obviously, if this interpretation is correct, popular culture can be considered one of the most important historical forces of this period.

Ware regards the airwomen as important figures in 1930s popular culture, along with movie stars and athletes. She devotes three pages to them, centering her comments around Earhart's career but also mentioning Helen Richey, Louise Thaden, and Jacqueline Cochran. Not only are they portrayed as important for their own difficult era — figures whose exploits could make people feel good despite bleak times — but they are presented as potential role models for several generations of American women.

Jeane Westin takes a very similar view of the thirties, acknowledging the overpowering reality of the Great Depression. In her mind, hard times called forth a wonderful array of adaptive if not heroic responses from hard-pressed American women in all walks of life. Her emphasis is less on celebrities or popular culture figures and more on ordinary women's coping strategies: "Why is it important to record their history? At the very least, today's women will feel a certain pride at being their daughters; at most, their insights will form a survival handbook for tomorrow" (1976, x). In short, because women of the thirties dealt with adversity so well, they should be admired and emulated.

Although Westin shows the courage and resilience of women of every status, she specifically acknowledges women who not only made do, but coped so well as to influence the world around them. They were "wonder women" and, like Brown's heroes of the twenties, they helped sound the death knell of the Victorian womanhood stereotype. Westin examines a broad array of people from Eleanor Roosevelt, to athlete Mildred "Babe" Didrikson, to Wallis Simpson. In this context, Westin too acknowledges the airwomen's accomplishments. Earhart is mentioned, but so are Ruth Nichols, Louise Thaden, Frances Marsalis, and Mae Haizlip in a generally focused two-paragraph entry. Unfortunately, the piece is so brief that it is not really clear why these women should be acknowledged. To compound matters, the material was poorly researched; for example, the names of the latter two pilots are misspelled (248).

Westin's work features, in addition to her own analysis, extensive documents in which the women of the thirties speak. Many extended quotations show coping strategies in the first person, some derived from published works and others from interviews that Westin herself conducted. Here she partially compensates for the less-than-exciting two paragraphs. The featured selection in the "Wonder Women" chapter is a three-page excerpt from Cochran's *The Stars at Noon*.

Just as the Great Depression dominated the thirties, World War II was the great phenomenon of the forties. It is widely believed, by scholars and casual observers, that the war set in motion social changes that greatly altered the lives of women. As we saw earlier, there is considerable truth to this view. But in *The Home Front and Beyond* (1982), Susan M. Hartmann notes that this decade harbored not only dynamic elements but also strong forces for continuity. Women did indeed take new steps, particularly in the workplace. They proved they could be fully competitive in many "masculine" spheres. Yet their admission to new roles was a matter of temporary military necessity. Many, both among the elite and the public at large, passionately believed in traditional family and gender roles and these beliefs underlay powerful imperatives for continuity. That women's wartime activities had expanded dramatically, everyone acknowledged and most applauded. That this expansion should permanently disrupt social patterns, on the other hand, was often anathema. While Hartmann, like most observers of women's history, is particularly interested in the potential for advancement, she believes that this dialogue between change and continuity defined the decade.

Hartmann identifies service in the military as one of the most significant departures from tradition. All of the armed services incorporated women; more than a third of a million women served. They were employed in virtually every function except combat. Moreover, since they had clearly proved their value, women became permanent parts of the military establishment. Although these developments were revolutionary, women in the services were not treated as the equal of men. Often, they were assigned routine "domestic" tasks. They were permitted special leave arrangements because of family obligations and in many cases were not permitted to supervise men. Moreover, official public relations statements sought to reassure the public that the servicewomen would not emerge any less feminine or less domestic than when they had entered. Indeed, service in the WACs was often portrayed as the performance of duties that women would normally undertake in civilian life (42).

Hartmann cites the WASP experience, both service and demobilization, as a prototype of the conflict between change and continuity that pervaded women's military service. She spends two and a half pages accurately describing the contributions of this group, as well as discussing the role of Love and Cochran. Although this discussion is brief, it does unquestionably identify airwomen as significant players in an important historical drama.

Despite occasional strong acknowledgment of the airwomen, can it really be said that writers in women's studies and women's history have kept the memory of the airwomen alive and vital? On balance, it does not appear that they have done so any more than the purveyors of aviation history. Certainly, women's biographical dictionaries withhold the adjective "notable" from the airwomen by the simple act of exclusion. Women in other fields are featured far more prominently. Such entries about airwomen as do appear are usually confined to Earhart; they are short, bland, and not always accurate.

Those narrative writers who concern themselves with broad developmental trends or abstract concepts almost completely ignore the airwomen. A focus on feminist activism, immediate economic and social conditions, and theory seems to have relegated the airwomen or their times to insignificance in the minds of these thinkers.

Authors who write directly of the time period in which the airwomen lived are only a bit more generous. They are much more willing to consider individuals important by virtue of their impact on popular culture or their function as role models, so it is not surprising that the airwomen are mentioned. Unfortunately, the treatments are disappointing in terms of both coverage and accuracy. One has the sense that the airwomen have been discovered, but that writers yet lack the ability to exploit their full potential.

Connections to Contemporary Female Pilots

Though historians and scholars, those who traditionally record the past and transmit the culture from generation to generation, have done little to proclaim the feats of the airwomen, the possible channels have not been exhausted. Have such standard chroniclers somehow been bypassed? Have legends of the airwomen entered contemporary consciousness in a less formal way, perhaps by word of mouth? Have the airwomen, despite everything, managed to serve as role models for today's successful female pilots? In at least one case, we can safely assume a strong interpersonal effect. One of today's female airline pilots, Terry London, is the daughter of WASP Barbara Erickson. Is there any evidence beyond this single, possibly idiosyncratic, case?

Relatively few contemporary female pilots have prepared narrative accounts, so it is difficult to inquire into their motivation. Bonnie Tiburzi, one of the first of the new generation of airline pilots, wrote the only full-length autobiography to date (1984). Though to some degree this book is a chronicle of feats accomplished and obstacles overcome, it features an excellent discussion of character development. A slight tendency toward self-congratulation is more than compensated for by a strong sense of mutual interdependence with regard to a number of other people. Tiburzi admits that early in her career she received much gratification from being the only woman in a masculine domain; she confesses that she felt resentment toward another female pilot who appeared at her airport. Some of the feelings she describes are reminiscent of Jacqueline Cochran, though the latter would never have admitted them. Although Tiburzi was also the first and for a time the only female pilot for American Airlines, she eventually came to see herself as an agent of social change if not a militant feminist. Indeed, she became a strong supporter of the International Society of Women Airline Pilots and accepted the need for collective action.

This work appears to be a good compilation of the important influences on Tiburzi's life and a straightforward indication of her basic thought processes. The airwomen do not appear to play a large role in either. To be sure, Tiburzi notes her early upbringing in a family with several male pilots. She thus became comfortable doing things with men and she came to prefer activities considered characteristically male. This often brought her into conflict with expectations for more feminine behavior. Tiburzi chafed under these gender roles, and in this context she briefly cites Amelia Earhart. This is a passing reference, however, merely to the effect that she regarded Earhart as a liberated woman (10–11). For all intents and purposes, Tiburzi came to her success as an airline pilot without benefit of the airwomen's memory.

Jeanna Yeager served as one of the two pilots on the spectacular 1986 nonstop, nonrefueled flight around the world. She and Dick Rutan flew the high-tech, composite-construction Voyager aircraft on a generally equatorial westbound route. Both the flight itself and the lengthy preparations had elements of high drama and the relationship between the two pilots promised to provide good human-interest stories. Accordingly the two published their adventures (Yeager and Rutan 1987).

Though the focus is on the present, there are brief chapters on the background of each. Unfortunately, Yeager is not very reflective and provides little explanation as to how she became successful and famous. In any event, it is clear that aviation was not part of her youth. There were no aviation personalities — either directly experienced or mediated by literature or legend — to influence her. Yeager appears to have been interested in horses since childhood, and she mentions a brief tenure as a high school track star.

At some point in her mid-twenties, Yeager professed an interest in learning to fly helicopters. This was apparently related in some way to a childhood fascination with dragonflies, whose structure and behavior she viewed as similar to that kind of aircraft. Preliminary training came to naught, and Yeager never did complete her helicopter rating. She was working at drafting and mechanical design when a series of chance meetings and job-related contacts led Yeager to the Rutan firm, which in turn was instrumental in the development of her career. The Rutan firm ultimately designed and built the Voyager airplane, and Yeager was selected as one of the pilots. Whatever the route that this woman followed to success in aviation, the airwomen were clearly not involved in paving the way.

Henry M. Holden and Lori Griffith (1991) conducted interviews with twenty-four female airline pilots; in their book *Ladybirds*, they present short sketches about each. Although they apparently did not explicitly probe for references to the airwomen, they did ask about early life and the events that had led each to a career in aviation. Most of these pilots attribute an interest in aviation — and the desire to make it a profession — to a family member, usually a father, or to some sort of singular experience. An unexpected, dramatic exposure to airplanes often inspired excitement so strong it led to a career. None of the twenty-four are reported to have mentioned the airwomen at all.

As far as the written words of today's female pilots are concerned, there is no acknowledgment of the airwomen. However, the written words are as yet few, and it behooves us to explore their motivation a bit more thoroughly. In late 1992, ten contemporary female pilots were interviewed by telephone specifically for this book. These ten were identified by the International Society of Women Airline Pilots, the worldwide organization of female pilots who fly for major carriers. Nine are employed in passenger service by major airlines; one suffered a layoff from a major airline but has accepted a position with another. All serve as cockpit crews in heavy jet airliners. It was not possible to construct a formal sample of the organization's membership, nor can it be claimed that membership is a perfect microcosm of the entire population of female airline pilots. Nonetheless, the individuals interviewed were diverse with respect to age, length of service, employer, and, as their interview responses showed, background and motivation.

The ten were asked to reflect on how and when they became interested in aviation and on what factors led to professional aspirations in this field. They were invited to list persons, events, or things they might have heard or read. Their answers were fascinating.

There appear to be several kinds of background experiences that direct contemporary women to successful careers as pilots. Respondents 1 and 2 came from families in which aviation was a part of everyday life. One was the daughter of an airline pilot who was also a leader in the design and construction of experimental aircraft. Because aviation was a large part of routine family activities, involvement in aviation and plans to become an aviation professional did not need any particular nurturing. This situation is very similar to Tiburzi's. The second respondent grew up in Alaska; regular commerce and local transportation were accomplished by plane. In addition, her father was an employee of the CAA. Again, aviation was simply a normal, prominent fact of life. These two pilots, while they certainly recognize the importance of their early lives and particularly their parents, stress their own reasoning or determination rather than role models. Early environment made aviation seem perfectly normal as a career path for respondents 1 and 2.

Respondents 3 and 4 also viewed their interest in aviation as something that did not require nurturing, but for different reasons. Respondent 3, trained in biology, had no particular interest in airplanes until she had completed college. Essentially, she just decided that flying looked like fun and quickly earned a private license. She found flying to be addictive and was progressively more drawn to it. Additional ratings and additional jobs followed, and eventually she ended up in regularly scheduled jets. Opportunity rather than conscious determination was the primary driver. Respondent 4, who never contemplated aviation until she was nineteen, was drawn to piloting more or less incidentally. She recognized the many professional advantages of an airline career and good timing opened an easy path for her. Logical decisions about salary, schedule, and advancement seem to have been the primary motivations. For important influences on their interests in aviation, these two cite early flight instructors. They seem in some sense similar to Yeager, who encountered some interesting situations and, applying inherent ability, went with the flow.

The remaining six respondents, however, were quite conscious that they had invaded a traditionally male field. Further, they understood that this had taken special motivation, which at least in part came from outside influences. Most often, this involved two things: an important aviation-related figure and strong parental encouragement of adventurism and independence. A family friend with an airplane, well remembered from childhood, was mentioned by several of the interviewees. Interestingly, in one case the family friend encouraged flying while the parents were horrified by the idea. Needless to say, the friend's influence eventually won out.

In at least four of these cases, female role models were an explicit part of the picture. Respondents 5 and 6 had actually become flight attendants as a result of their childhood fascination with airplanes. Initially they had received the message that the appropriate niche for women in aviation was as a servant in the airliner cabin. These two, however, were deeply impressed by the very first female airline pilots on whose flights they sometimes served. Not only were these pilots role models to be admired from half a fuselage-length away, but they took the time to encourage the ambitious flight attendants to seek new careers. After much hard work and training, both moved to the cockpit. In the case of respondent 5, this was after thirteen years of serving rubber chicken.

Respondent 7's situation is very similar. Childhood fascination with airplanes developed into admiration for flight attendants and plans to be one. However, since she was younger than respondents 5 and 6, when she began serious career planning there were already several female airline pilots; respondent 7 was able to switch role models. Her father, who had for years suffered the frustration of not being able to fly, urged her onward.

Respondent 8 grew up near an airport and also had a life-long love affair with aviation. For her and one other respondent, the first landing on the moon was a dramatic reinforcer of aviation interest. This affection for flying was translated into professional ambition when respondent 8 encountered, at college, a woman who piloted the local parachutists' jump plane.

Respondent 9 remembered a dashing family friend who was also a pilot, but her only female role model was her mother. The latter warmly encouraged her choice of this nontraditional career. Respondent 10 remembers the Civil Air Patrol as an important part of her childhood, but recalls no feminine form among those who helped pave her way.

What does all of this tell us? At least six of these female pilots, respondents 5 through 10, specifically acknowledged significant role models in their development of an aviation career. In addition, although respondents 1 and 2 did not speak in terms of role models, their youthful backgrounds, filled with aviation experiences, had some important effects. It is especially interesting, of course, that many of these role models were women, often pilots. In a word, the image of female pilots can be an important contribution to the development of professional careers.

Given this, it should come as no surprise that these pilots, conscious of their professional success and the factors that helped them achieve it, themselves function as role models for younger women. Asked to discuss their efforts to educate or serve younger women seeking a career in aviation, the respondents listed a vast array of activities. These included appearances at high school and college career fairs, participation in community service programs sponsored by their employers, service as a one-on-one career counsellor at various institutions, service as an adviser to various girls' groups such as the Girl Scouts, and various kinds of service for scholarship programs serving women. Five of the ten are heavily involved in such activities while two have at least some involvement. Even though three do not do things of this kind, the collective record of service is outstanding by any standard.

One might think that if the memory of the airwomen had been kept alive in our culture, these remarkable women would be part of the communication chain. This does not, however, seem to be the case. Throughout the discussion of the development of aviation and professional interests, there was no mention of any airwoman. In this respect the respondents are similar to those pilots interviewed by Holden and Griffith (1991).

However, the airwomen are not totally disconnected from contemporary female pilots. In another part of the interview, the ten respondents were presented with the names of twenty-four people, places, or things, most relating to aviation. They were asked to indicate whether they had ever heard of each. Embedded in the list were nine items: the names of six airwomen, the Air Transport Auxiliary, the WASP, and Beryl Markham's *West With the Night*. In order to make the exercise seem less like a test, each respondent was asked to indicate a positive, negative, or neutral view toward each item with which she was familiar. Of course, it would have been possible for respondents to falsely indicate familiarity simply by claiming it. If the respondents feared they were being judged on the basis of their knowledge, there might have been a tendency to lie in this way. To guard against this, five totally spurious but plausible items were put into the list (for example, names like Emma Colton); no one could be familiar with them because they did not exist. If a respondent were inclined to lie, she would claim knowledge of these five items; if she were being truthful, she would admit no knowledge of these five. Since there were five items and ten respondents, there were fifty opportunities for false claims. Only one occurred, indicating that there was no problem with the respondents' honesty.

On average, the respondents were familiar with fewer than six of the nine items. While two were familiar with all nine, two others claimed to have heard of only two. As Table 4.1 indicates, knowledge varies according to the item.

Table 4.1
Respondents' Familiarity with Nine Items

Item	Number of Respondents Familiar with Item
acronym "WASP"	10
Jacqueline Cochran	9
West With the Night	9
Helen Richey	7
Harriet Quimby	6
Louise Thaden	5
Katherine Stinson	5
Air Transport Auxiliary	4
Nancy Harkness Love	3

The respondents knew about the WASPs and Jacqueline Cochran; they had read *West With the Night*. The other items are less familiar. To be sure, these women are considerably more knowledgeable than the general public, where the average number of items known would have been close to zero. However, the overall level of familiarity does not suggest a transmission line of any broader gauge than that provided by aviation historians or scholars in women's studies. It is perhaps not surprising that none of the respondents claims, in the course of her professional development, to have run up a debt to the airwomen.

Conclusion

The memory of the airwomen remains indistinct and shadowed, in contrast to the glittering public image of their own time. Even somber observers, who do not have to sell their words in tomorrow's edition and who have the vantage of time, have not restored the luster. Aviation historians are at least marginally aware of the airwomen, but their treatments are often brief, less than completely accurate, and vague in purpose. They present no clear picture of how the airwomen fit into the big picture of the developing technology of aviation. Some authors emphasize an unfortunate theme that has been around since the late twenties: they pay more attention to early female flight attendants than to early female pilots.

More informal treatments of aviation history have many of the same flaws. Though the airwomen generated a great many adventure and human-interest stories — the very stuff of this kind of writing — authors have let these fascinating stories lie dormant. The airwomen remain very much in the background. A bit of Earhart, some photos of stewardesses, and women in early aviation have been covered.

By its nature, much historical commentary has a strong conservative bent; but appropriately, this conservatism is from time to time challenged by alternate interpretations. Such a challenge is currently at hand in the form of women's history and women's studies. Careful and responsible observers are providing a different view. Given their greater interest in, for example, popular culture, one might expect that these observers would cast the airwomen in a different light. In general, however, scholars who focus specifically on women of the airwomen's day are not much more generous than their traditional counterparts.

Finally, a series of interviews with female airline pilots revealed little evidence that the memory of the airwomen was somehow maintained through communications processes that did not depend on the print media. Attention for the airwomen remains elusive. At the conclusion of the last chapter, it was clear that following 1944, the airwomen's memory lost an important battle. At the conclusion of this one, it appears the war was lost as well. However, aviation history is relatively new and undeveloped. As it grows and matures, it is certainly possible that more complete and more accurate pictures of the airwomen will emerge. Valerie Moolman's *Women Aloft* (1981) is a step in the right direction.

Women's history is similarly an area of study with only a brief track record. Its broad, philosophical base seems congenial to the airwomen, and it seems that some authors are willing to consider their importance even if they have as yet poorly implemented that consideration. An occasional title — for example, Angela Howard Zophy's *Handbook of American Women's History* (1990) — looks almost encouraging. As scholars have an opportunity to consider broader horizons, perhaps the airwomen will be reevaluated.

Although contemporary female airline pilots did not, apparently, take inspiration from the airwomen, they are not totally ignorant of them. Indeed, a few have made it their business to know them well. This familiarity, limited though it is, could be important. These contemporary pilots have a clear commitment to serve as role models for those younger still. Equipped as they are with some knowledge, the airwomen just might be part of their message. Nonetheless, the memory of the airwomen has a long way to go.

❧ 5 ❧
Explanations:
Promise Unrealized

Despite a few bright spots, the message of the last two chapters is bleak indeed. Why did the airwomen not leave a greater legacy? Rarely, if ever, is there a single cause for anything important. As history unfolds, one can at best hope to identify a number of contributing factors, necessary conditions for particular events. Certainly the rich history of the airwomen resists any simple explanation. Moreover, in addition to multiple causes, there are usually multiple effects. Many writers and historians have reported on, for example, the causes of the Civil War. As we know from any decent high school history class, there were many, having to do with geography, economics, international relations, and particular personalities, to name but a few. The Civil War itself was a massive, multifaceted event. In addition to celebrated battlefield engagements, it involved military occupation, a total restructuring of partisan politics, technological innovations, and a radical disruption of production. When one asks about the causes of the Civil War, one must also ask what it was that was caused; much more than a single event needs to be explained.

Analyzing the *absence* of events, such as the lack of an airwomen's legacy, adds the confounding dimension of imagining what might have been. This requires a hypothetical but plausible alternative to what actually happened, and makes the task all the more daunting. The purpose of this chapter, then, is to identify some probable causes of the airwomen's fall to obscurity, particularly those causes that are likely to have counter-

183

parts in the present. The goal is not to find a master solution, as in a great mystery novel, but instead to fit a few clues together in order to provide insight into today's world.

It may be tempting simply to attribute the lack of an airwomen's legacy to the fact that our culture is basically sexist. However, it is also important to inquire how the airwomen might have succeeded in spite of that prejudice, and how their success might have caused a decline in that prejudice. Previous chapters have detailed many examples of male prejudice against the airwomen. These, however, are viewed as targets, while the airwomen's impressive record is the metaphorical equivalent of a gun. The missed shots are explained not in terms of how hard the targets were to be hit, but in terms of how the aim might have been improved.

Does such an approach "blame the victim?" As stated at the outset, this book is not an exercise in hero-worship. It is not particularly useful to assume that the airwomen could do no wrong and that they must be defended at all costs. It is far more productive to try, with the luxury of all our current experience, to imagine better outcomes. What factors under the airwomen's control might have contributed to such outcomes? It is not the intent to disparage those airwomen who made what may appear, in retrospect, to be bad decisions, but to see whether there are any lessons for more effective strategies in present or future situations where women must confront a sexist world. Improvement can never occur without critical examination of what has gone before. Such examination does not blame people for acts that may have been mistakes, but instead honors those who had the courage to take on huge challenges.

This book has already identified several factors that harmed the cause of the airwomen. A prominent example is the development of a unique women's role in aviation, that of stewardess. This role was so compelling that even observers who were very favorable to the airwomen gave it a good deal of attention. Also important is the very individualistic, self-centered posture adopted by a number of female pilots in the post-1944 decade. They cared little for anything but personal success; certainly the general cause of women in aviation was far from their minds. As a final example, the postwar Ninety-Nines failed to grasp that it was time to begin serving the job-related needs of its constituents. Instead, that organization tried to re-create a past in which the needs of female pilots involved demonstration of competence rather than professional aspirations. Thus, major themes have already been introduced; this chapter will consider some of the more important ones in a more formal and complete manner.

A Bigger and Better Record?

The airwomen's accomplishments, it is clear, were dramatic indeed. It appears, however, that they were not great enough to reach the notice of the public, historians, or future activists. This suggests that, had the airwomen posted even grander achievements, recognition would have been more complete and lasting.

On its face, this is not an unreasonable expectation. Of course, we cannot know where that attention threshold might have been: would twice as much accomplishment have triggered the development of a legacy? In any case, it seems likely that a better record would have received greater media coverage. Chapter 2 demonstrated that the airwomen by and large enjoyed very positive attention from journalists and writers. There is nothing to suggest that additional airwomen stories would have exceeded the public appetite for them.

Public attention spans today are notoriously short; the media dart from one issue to another with astonishing frequency. Television in particular appears to have induced a need for what communications experts call "stimulus variability." No series of events can hold center stage for long; the public tires of a given topic and demands to move on to another. Even great crises and huge scandals are here today, gone tomorrow. In the era of the airwomen, this was not the case. Female pilots were highly newsworthy for over fifteen years. Had there been more to report about them, it would have been reported. More attention, in turn, might have improved the prospects for lasting impact.

Why did the airwomen not generate a better record? This question can be approached from two closely related standpoints: first, external forces may have prevented them from doing so; second, the airwomen themselves may not have been sufficiently motivated.

One's immediate inclination is to emphasize the first. It is clear that the airwomen not only pushed back many frontiers, but were also anxious to take on additional and more difficult challenges. For example, both Alys McKey and Ruth Law, among the earliest of the airwomen, sought to enter World War I as combat pilots. Naturally, at that time, they were turned down. In the early part of the twentieth century, women were thought to need special help and protection. Their participation in any kind of aviation was viewed dubiously in some quarters on account of the perceived dangers. Combat risks were unthinkable. Women's interest in aerial combat roles did not diminish, however, as evidenced by Louise Thaden's story about female warplane pilots published in 1938. By the time of World War II, traditional fears had not much diminished. The "need" to protect women from actual fighting was involved in the decision to deny the WASP a role in long-distance, trans-oceanic ferrying. Indeed, despite the intervening years, female pilots are still, in the 1990s, not permitted to fly combat missions. It would be unfair to charge the air-women of the great wars with inadequate drive to get into military combat roles. Their great-granddaughters, with the advantages of more experience and a slightly changed culture, still have not accomplished it. Particularly in this area, tradition has been a monster obstacle.

Another external factor limiting the airwomen may have been the equipment available to them. Many have claimed that men got the newest and most advanced airplanes, while the airwomen had to settle for second-line machines or hand-me-downs. Airplanes have always been costly and the airwomen, like everyone else, had to struggle to afford appropriate machines. Often they did not have the best. Ruth Law's Curtiss Pusher was not quite an antique but certainly it was obsolescent at the time of her record flight from Chicago to Elmira (Oakes 1978, 40). Bobbi Trout was constantly at odds with her equipment supplier; she and Hugh Bone, builder of the Golden Eagle aircraft line, apparently had very different views on which records to challenge and how to challenge them (Veca and Mazio 1987, 133–206). The question, of course, is to what degree women were systematically denied top-notch equipment because of their gender and to what degree their problems were simply part of the phenomenon of appetite exceeding budget that most pilots experience.

Once separate races and record categories were established for women, equipment discrimination was easier to practice. The creation of these separate categories greatly expanded the number of newsworthy

competitions and thereby assured considerable attention for the airwomen. At the same time, however, separate categories are rarely considered equal categories; there was certainly an assumption that women's
aerial performance would not be at the same level as that of their male
counterparts. Many airwomen recognized that separate categories of competition reinforced discrimination, but judged the increased exposure and
publicity worth the cost. Common acceptance of separateness thus guaranteed that lesser aircraft could win in the women's events. Budget-con-
scious manufacturers accordingly made lesser aircraft available. A more
direct consideration of separate competitive events appears later in this
chapter.

Some airwomen staunchly believed that equipment discrimination
was deeper than this. Manufacturers, they argued, deliberately kept them
away from the highest-performance airplanes out of concern that their
flying skills were not equal to the challenge of sophisticated machines.
The manufacturers presumably feared that their airplanes would be damaged or that bad publicity would follow any mishap involving a woman.

Elinor Smith complained bitterly about some of the old and tired
equipment available for one of her endurance flights, particularly a notoriously unreliable Curtiss Carrier Pigeon refueler (E. Smith 1981, 126).
She also believed that women were disadvantaged by lack of access to
military aircraft. At the same time, Smith became a protégé of Giuseppe
Bellanca, one of the leading manufacturers of the twenties and thirties.
Bellanca provided state-of-the-art machinery to Smith, especially for her
altitude record attempts. Indeed, these same attempts got Smith access to
other important equipment. The U.S. Navy had become very interested
in high-altitude experimentation. It had built a lot of test equipment,
including a radically innovative altitude chamber. This large tank simulated the low temperature, low pressure, and low oxygen concentrations
found at various extreme heights above the earth. Pilots were sealed into
this chamber in order to train for the grueling conditions of actual flight.
All this was made available to Smith before she challenged the strato-
sphere (E. Smith 1981, 200–217).

Both Louise Thaden and Amelia Earhart also believed that women
had chronic problems with equipment quality. In some respects this is
curious, for Thaden flew the newest and best of Travel Air and Beechcraft

airplanes. Indeed, Walter Beech tempted her back into active racing with the offer of a powerful new retractable landing gear machine. Similarly, Earhart was the frequent beneficiary of G. P. Putnam's personal largesse and his skill in raising money for airplane purchases. Apparently, both thought that scarcity of good airplanes was a problem for women in general, not necessarily one from which they personally suffered (Thaden 1938, 273). It is true that Earhart herself was once denied an airplane. She wanted a new cabin monoplane for the 1929 Women's Air Derby, but Giuseppe Bellanca refused to sell her one (Lovell 1989, 143). It is unlikely that this was an act of gender discrimination, given the way this manufacturer had promoted Smith; both Smith and Earhart's biographer argue that Bellanca did not think Earhart could fly well enough to avoid danger. In any case, Earhart was able to obtain a Lockheed Vega for this event and retained this top-of-the-line type for several of her record attempts. Later, with the support of a major research university, Earhart obtained an advanced, twin-engine Lockheed for her around-the-world attempt.

Ruth Nichols was also able to secure the enthusiastic backing of patrons. For money and technical support, she relied on Clarence Chamberlain, a well-known early aviation entrepreneur. The aircraft she used for her most notable feats was also a Lockheed Vega, but this one was owned by the Crosley Radio Corporation, a high-tech business giant of the thirties (Nichols 1957, 229). Powell Crosley, president and CEO, allowed Nichols to use the plane through a series of record attempts, crashes, and rebuilds until, ultimately, it was totaled.

Many of the less well-known airwomen were also able to obtain first-rate aircraft. Pancho Barnes raced in a Travel Air Mystery Ship (Tate 1984, 67). Phoebe Omlie had an arrangement much like that of Louise Thaden with Walter Beech, though hers was with Mono Aircraft Corporation, builders of the Monocoupe line of racers. Mae Haizlip, a leading money winner among race pilots, flew first-line Granville Gee Bee, Laird, and Wedell-Williams planes (Oakes 1985, 38).

Well-connected Jacqueline Cochran never had trouble obtaining the airplanes she needed, having at various times raced custom-built Gee Bees, Northrop Gammas, and Beechcraft Staggerwings. Her ultimate triumph, winning the Bendix transcontinental race, followed her ultimate acquisition. Far from denying a woman access to his newest and most sophisticated design, Alexander de Seversky specifically sought Cochran out to fly his prototype fighter.

The evidence, thus, is rather mixed. While most manufacturers accepted the assumption that women's competitions would be less rigorous, they did not always withhold the best planes from the airwomen. This is not to argue that aircraft manufacturing firms, always controlled by men, were necessarily motivated by high ideals of gender equality. Rather, it may be that corporations were anxious for the attention that female pilots, relatively novel and therefore newsworthy, could draw to their products. Though not particularly enlightened, this strategy, based on the true premise that female pilots were very able indeed, was more effective than that of denying airplanes to women, based as it was on the false premise that they were not.

One of the most curious facts about the history of the airwomen is that many of them appeared to quit in midstride, to step down from a brilliant career at a young age. This withdrawal had profound implications for the sheer magnitude of the airwomen's record. Ruth Law, Katherine Stinson, Ruth Nichols, Louise Thaden, and the WASP came to the most sudden stops. Had they continued, there seems little doubt that collective accomplishments would have been even greater.

Considering that these airwomen might not have withdrawn, is it reasonable to assign them some responsibility for not staying the course? Or are external causes once again implicated? Undoubtedly, both kinds of factors contributed. Mathilde Moisant, one of the earliest airwomen, quit abruptly following a fiery crash that very nearly claimed her life; a more explicit external influence can hardly be imagined. On the other hand, late in life she told a Columbia University Oral History Project reporter that she left aviation because she did not wish to cope with the pervasive misogyny in the aviation community (S. Harris 1970, 240). Other cases are also ambiguous. As noted earlier, Ruth Law discovered that she had retired by reading an article in the newspaper; her husband simply forbade her to continue her flying career and announced as much to the press. Law was distressed by this and suffered greatly as a result. Who is to blame: her husband for his outrageous behavior, or Law for her failure to object? If she had defied her husband, she might well have become a role model for future generations. She might have wanted to defy him, but could she actually have done so given the times and the conditions?

Katherine Stinson fought vigorously to establish herself as a pilot and clearly believed that women in general should have a role in aviation. Like Law and McKey, she tried to fight in World War I. While that attempt failed, her next exercise in perseverance paid off. In 1920 she was appointed as a pilot in the U.S. Airmail Service, a very prestigious position. Thus there was a thoroughly successful woman among the first really coherent group of civilian professional pilots in the United States. But Stinson resigned after only a few flights. She gave no reasons, but in 1918 she had contracted a serious illness in France; it is possible that she was simply too ill to continue the rigors of open-cockpit pilotage.

Nichols's flying career also ended abruptly, though she continued, not altogether successfully, to promote aviation for many years. More than any other airwoman, Nichols was a veteran of many aviation accidents. She had given much blood and broken many bones for the cause. Nichols was certainly not ready for the scrap heap — though her Lockheed Vega mentioned a few paragraphs above unquestionably was. But she had been badly battered and had narrowly escaped death many times. It is sad that Nichols did not continue to demonstrate skill and accomplishment, but it seems unreasonable to reproach her for not having made still more sacrifices.

Thaden retired and then returned to flying several times. Her writings make it clear that the periods of inactivity were difficult and unsatisfying. Nonetheless, she abandoned her program of racing and record setting, and perhaps more tragically, her professional career with Beechcraft. Given her successes, her clear desire to continue, and her sensitivity to the overall status of women in the culture, it is difficult to explain her departure. Thaden's domestic situation was a fairly traditional one and her husband also had an aeronautical career. Whether his agenda and hers were in conflict is not known, but they did not always fit together well; Thaden as race pilot and husband as designer of the planes was not a successful combination. Thaden did not reject her family role, but did manage to work on and off for a manufacturer. She also served in the Air Marking Program, despite her husband and children. Her record of liberation was, for the times, not at all shabby.

Most WASPs retired in 1944, and for most this was a forced retirement of the most overt kind. To be sure, throughout the years of active service, some military airwomen did resign their positions. For the most part, this was due to family considerations. The departure rate was not excessive, however, and no one has criticized the WASPs for failure to follow through. At deactivation, many WASPs simply went home, never to exhibit much interest in a professional flying career again. It is abundantly clear, however, that most WASPs would have been delighted to continue as pilots but were told in no uncertain terms that they were not wanted. Insufficient motivation was not a factor with this group of airwomen.

Other airwomen also withdrew, leaving behind little explanation. Less famous women such as Phoebe Omlie and Bobbi Trout left no notes or letters to explain their disappearance. Thus, the reason so many airwomen chose to withdraw rather than continue to build an impressive record remains obscure. While there are cases in which motivation seems to have been lacking, there are others in which the airwomen were clearly constrained — and still others in which airwomen really did a great deal. To be sure, today's successful professional women will be able to show accomplishments in many ways more dramatic. But no one should diminish the airwomen by comparison. The system of constraints under which they lived was different. If the general maxim for the achievement of full citizenship for women is to achieve as much as possible, the airwomen did not fall greatly short of this goal. A bigger and better record would have been desirable, to be sure, but most of the impediments to the creation of such a record were externally imposed.

If Women Can Do It, Anyone Can: Economics and Emotion

Previous chapters of this book have discussed the notion that "if women can do it, anyone can." This theme was exploited to encourage wider public participation in aviation. Indeed, many early professional opportunities for female pilots were generated by precisely this mechanism. This was particularly important in the twenties, as aviation began to mature. The commercial potential of aviation, entrepreneurs realized, could not be developed if the prevailing image of aviation endured. There thus developed a substantial economic demand for female pilots.

Corn (1983, 71–91) points out some extremely important features of this business arrangement. Most important, the female pilots in this role had to be as conventional as possible. As pilots they were inherently unconventional, there was strong pressure to make them conform to the typical image of American women in other ways. If they were perceived to be unfeminine or of dramatically unusual ability, the rationale for employing them in the first place would be lost. They were thus encouraged to champion family values and project a "ladylike" appearance. In a very real sense they had to appear less competent than they really were.

Many of the airwomen were aware that they were being exploited in this way. Despite the feminism that many of them espoused, there were powerful motivations to accept such treatment. Most important, they understood the big picture; they, too, wanted aviation to thrive. Their motivation was thus more emotional, more that of true believer in "the winged gospel," than that of their profit-minded employers, but they nonetheless came out at the same place.

Corn believes that the strategy, "If women can do it, anyone can," succeeded grandly. Aviation did change its image and commercial applications have mushroomed from the thirties to the present day. Women, exploited in the process, were crucial. "Domesticating" aviation helped to sell it; this is especially apparent in the popularity of flying married couples such as the Omlies, the Haizlips, the O'Donnells, and especially the Lindberghs.

For women, the short-term benefits of this arrangement were increased professional opportunities. In the long term, however, the deal seems much less advantageous. "Domestication" of aviation involved their separation from the aeronautical mainstream: women had a special role, and one that was very constrained.

Moreover, the domesticating of aviation was, by its nature, temporary. By the late thirties, the public had generally accepted aviation as an everyday part of modern life and business. Airmail, passenger service, and other air commerce were well established. Although female pilots were not the only agents bringing about this change, they had done their job well. Because of their success, symbols of the domestication of aviation were no longer necessary. The point had been made. To put the matter only slightly too crudely, the airwomen had worked themselves out of a job. According to this interpretation, there is no mystery or difficulty about the failure of a women pilots' professional stage to develop; it was simple economics.

While there is considerable truth in this view, exploitation of the theme "If women can do it, anyone can," was not completely over by the late thirties. Indeed, its clumsy and insensitive use played some part in the tragic history of the WASPs. Much of the discussion of that organization was straightforward exchange about the availability of human resources. Indeed, the very creation of the WASPs was justified by an anticipated shortage of male pilots. In general, military planners were interested in the plan according to the severity of their need to move airplanes around. General Tunner, among others, was looking for a tool to help him get his job done. Similar motivation was shown by many who originally objected to the WASPs, but were converted into supporters when they saw that performance was indeed at a high level. Even some of the discussion surrounding the disbandment of the WASPs was conducted in similar personnel management terms; some observers believed that continued training of any type of pilots was no longer necessary.

However, opposition to the WASPs, and often support for that matter, often hinged on emotion as well as human resources calculations. Bitter anti-WASPs diatribes were common. Many base commanders detested the idea of female pilots and went to great lengths to deny WASPs any active role. There is even evidence that planes were sabotaged and at least one death may have resulted (Douglas 1991, 51). Such hateful and violent actions did not stem from a careful assessment of performance.

This emotional hostility can be attributed, in part, to the sexism that permeated the entire culture. It is also true, however, that key leaders, often male generals, paid too little attention to the emotional responses that the activities of the WASPs might generate. While General Arnold, top officer of the Army Air Forces, generally supported the WASP program, he did have some peculiar uses in mind for it. As noted earlier, he employed WASPs to fly demonstrations of the P-39, a fighter plane with a reputation for being intractable, and the B-26, a high-performance medium bomber that had a relatively high wing loading, in order to cajole men into accepting these machines. However, many male pilots saw this for what it was: an attempt to coerce them through shame (Douglas 1991, 55). The WASPs' special mission thus generated resentment and made no allies for the airwomen.

There were parallel mistakes in other uses of the WASPs. Many civilian promoters in the twenties and thirties hired female pilots for the simple reason that they attracted attention. World War II military brass repeated this stunt in promoting the WASPs as part of a campaign of patriotic boosterism. Physically attractive women selflessly serving their country got the attention of the media. While there probably were some public morale benefits from this, there was a downside: the WASPs received recognition and glory while their male counterparts, performing the same function, got no press at all. These men, understandably if not justifiably, were resentful of the WASPs' publicity. This public relations ploy was a disaster for the airwomen because it angered an important segment of the flying community.

The role implicit in the notion, "If women can fly, anyone can," separated women from the aviation mainstream. It had no long-term career implications since success meant working oneself out of a job. To add insult to economic injury, this notion gained the WASPs resentment and hostility from parts of the aviation community. It was extremely difficult for the airwomen to avoid being used in this way. They understood that their role in "domesticating" aviation was good for the society as a whole. Moreover, by agreeing to this arrangement the airwomen were responding to an economic demand. A professionally ambitious group could hardly have done otherwise. The choice for many was these positions or none. It is certainly not at all clear that choosing none would have better served the cause of women in aviation.

The Doctrine of Separateness

As the Supreme Court argued in *Brown v. Board of Education*, separateness, as it results from racially segregated education, is inherently unequal. The same might be said about gender separation in the time of the airwomen. As Chapter 2 demonstrates, the earliest airwomen did not have to contend with the issue of a separate role. While the culture did not treat them the same as men — they were denied a role in World War I, for example, and Katherine Stinson had to fight for a position as an airmail pilot — neither were they directed toward distinct aviation roles. It was not until well into the twenties that women began to be treated as a separate category of aviation participants. Concerns about fashion and family responsibilities appeared in the press; female pilots were increasingly used as symbols of domestication; eventually, the role of stewardess was advanced as ideal for women.

The airwomen had little control over most of this, but there was, as we have seen, a certain amount of acquiescence. Women were subject to equipment discrimination because manufacturers assumed that lesser airplanes could win competitive events restricted to women, but it was women themselves who argued for special record categories, for unique women's prizes, and for separate race competitions. When Viola Gentry completed the first women's endurance flight in 1928, it was not officially recognized by the Federation Aeronautique Internationale (FAI), the worldwide recorder of aviation records; the flight was not a record in terms of absolute duration — men had already stayed aloft longer — and there was no separate category for women's efforts (Gentry 1975, 75). Recognition for this flight and all previous accomplishments of the airwomen came only in the press and the mind of the public. Many people were not happy with this state of affairs, least of all Gentry. She had official observers from the FAI present as part of an organized campaign to get the FAI to establish separate categories for women. Success came quickly as such categories were established in June of 1929. Indeed, since the whole idea was consistent with domesticizing aviation, the only opposition seems to have come from bureaucratic inertia. The airwomen, by and large, were delighted. Indeed, even contemporary writers rhapsodize over this decision: previously, female pilots "did not receive the recognition they deserved for their flights. . . . Now, women's achievements could be seen in a perspective, and they would be judged on the basis of progress they were making" (Brooks-Pazmany 1983, 34). It is difficult to understand, however, what "perspective" and "progress" were supposed to mean.

Women's air racing has a somewhat different history, but many similar dimensions. In the thirties there were a number of racing events in which men and women competed with one another, most prominently the Bendix. However, such open competitions were controversial, and in certain places, at certain times, women were prohibited from competing. Perhaps in response to this occasional exclusion, there were a large number of women-only events — and even in 1934 a Women's National Air Meet — actively sponsored and promoted by women. In the thirties women raced for, among other awards, the Amelia Earhart Trophy, the Joyce Hartung Trophy, and the Ruth Chatterton Trophy, all named after their sponsors (Oakes 1985, 44–45). Earhart went so far as to commission special awards for the best girl contestants in national model airplane contests (Corn 1983, 118).

Should the airwomen have eschewed separate competition categories in order to strengthen the claim that they were as good as any in the business? Elinor Smith argued that women should not, by and large, compete in unlimited fashion with men. Her concern was that women as a group were less experienced and such efforts would court danger. As noted earlier, the main argument in favor of separate categories centered around the increased attention that special official categories conferred on female pilots. Had the airwomen not opted for women-only competitions, they undoubtedly would have won a few more open races and posted a few more absolute records. This would have established them as anyone's equal, but given the relatively small number of airwomen, fewer of them would have achieved public prominence. It is difficult to ascertain which is the better ingredient for a legacy: a small number of supercompetitive airwomen, or a larger number of highly competent but still very well-publicized airwomen.

Another thrust toward separatism was the association of female pilots with domestic responsibilities, which have traditionally been viewed as women's main priority. Airwomen's commitment to aviation, in the minds of many, had to be limited by their commitment to more basic social mores. While this notion was deeply rooted in the culture and thus, in a sense, externally imposed, a large number of airwomen did embrace traditional values. Many were quite comfortable with the idea that they were different from male pilots because of family obligations or other domestic concerns.

This seeming inconsistency is well illustrated by the mixed messages communicated by the Ninety-Nines. By the early thirties, that organization was well established and began to publish a monthly magazine that went by the titles *99er, Airwoman,* and *Ninety-Nine News* at various times. Although their degree of involvement varied, most prominent airwomen belonged to this organization. It is not surprising, therefore, that its publications should contain some fairly aggressive advocacy for female pilots and that they should gain favorable comment from, for example, the American Association of University Women. The monthly reported on cases of suspected professional discrimination by businesses or government and took some very pointed shots at men and their sexist views and language. *Airwoman* featured a column, "Bread and Butter Flying," that was an explicit attempt to serve women who flew professionally.

However, the publication also had some content that was less forward-looking. Very traditional women's concerns received much attention in columns on such topics as cooking. To be sure, there is nothing wrong with articles on food in aviation publications, particularly if, as was often the case in these magazines, they discuss the favorite dishes of pilots regardless of gender. But the Ninety-Nines's food editor assumed that her readers had heavy kitchen responsibilities, entering into discussions, for example, of what is likely to be gastronomically pleasing to men. To be sure, these columns were different from ones appearing in other women's magazines in that they urged, for example, the baking of cakes shaped like airplanes, but they were solidly domestic nonetheless.

These monthlies likewise had a fashion section, which discussed not only what the well-dressed female pilot would wear, but also children's clothes. One story starts from the assumption that women need large amounts of clothing to be properly dressed for all occasions; it then proceeds to discuss the advantages, for women, of airplanes with large baggage compartments! Tradition reaches a poignant if inappropriate peak in a 1935 article about Wiley Post and Will Rogers, recently dead in an air crash in Alaska. The focus was on the wives of these two famous aviators: "Theirs has been the rare privilege of exercising woman's great possessiveness over truly noble men. What more can any woman ask?" ("Alaska Crash," 1935).

Of course, it is impossible for any group to suddenly or completely shed expectations born of venerable tradition. At least some part of the behavior of all human beings stems from tradition, and it would probably be undesirable if it did not. The crucial question is how extensive tradition's influence is, how inimical it is to real advance. In the case of the airwomen, this is difficult to assess, but airwomen did frequently capitulate to the demands of family rather than continue in a promising career. In at least some of the cases, it seems regrettable indeed.

Perhaps the most extreme example is that of Neta Snook Southern, a very successful airwoman who taught Amelia Earhart to fly. Southern bought a surplus Jenny at the end of World War I and proceeded to tour the barnstorming circuit. Successful, but tiring of this lifestyle, she migrated to California and established a business to offer flying instruction. She did very well at this, gaining, rather incidentally, Earhart as one of her clients. Her reputation for safety and caution did not rub off on Earhart, but did gain her some employment in technical and manufacturing subcontracting.

Southern eventually married, and childbearing had a massive impact on her involvement in aviation. For reasons that she does not explain, she vowed that if she had a healthy baby, she would give up flying forever. The arrival of the child confirmed this decision and she ended her career. Southern never flew again, even as a passenger. Indeed, when asked to join the Ninety-Nines, she refused, offering as an excuse that her son's welfare would not be served by a revival of her aviation interests (Southern 1974, 157). Such subordination to traditional expectations is difficult to understand in one who travelled from Chicago to Newport News, the site of her flying school, by hopping freight trains, and who ran a successful flying service in her early twenties.

Of course, it is possible to reconcile family roles with professional ones. And it is not necessary to refer to contemporary successful women to prove the point. Betty Huyler Gillies was very domestically inclined when she first learned to fly: "I took up flying for one of the oldest reasons known to womankind — to catch my man!" (Gillies 1970). However strong this matrimonial motivation may have been, it did not prevent her from becoming a prominent WASP and later a successful test pilot. Indeed, it was she who formulated the dramatic offer of professional service to the government following the disbanding of December 1944.

A third element in the airwomen's separatism involved the attempt to create exclusive preserves. Certain types of aviation activity, often involving some similarity to traditional female responsibilities, were thought especially appropriate for women. Beginning in the late twenties, journalists, educators, entrepreneurs, and others attempted to define a unique aviation role for women. Indeed, women themselves often sought to establish a distinctive position in the flying community and tried to lay claim to an exclusive turf. If one views women's involvement in aviation in economic terms, that is, strictly as a matter of competition with previously established men, then seeking out a particular market niche, one not presently occupied, seems to make good sense. There is, however, a distinct problem with such an approach: it involves an admission that there are male arenas in which the airwomen cannot compete.

The Ninety-Nines very explicitly joined the search for a women's niche. Its attempt to define women's substantive aviation areas was not very forward-looking. For example, Barbara Southgate's "Not in Competition With Men" (1934) reads less like an aggressive market-inspired strategy than an attempt to reassure readers that airwomen represent no threat. She advises that there are "domains" already controlled by male pilots, including, for example, transport flying, that women would be wise to eschew. Southgate also wanted to establish strictly female counter-domains, the most prominent of which was to be the education of the masses, especially children, to the wonders of aviation: education, child care, and the domesticization of aviation all rolled into one. The only thing lacking in this article was safety, a topic that others had suggested should be a female aviation preserve. It is not clear that this approach soothed any male anxieties about aggressive airwomen. In fact, it is more likely that the female readers of such material were encouraged to limit their ambition — and once again, to appear less competent than they really were.

Other women advanced similar themes. In the late thirties, an organization called the Women Flyers of America (WFA) attracted a great deal of attention. It was essentially a national flying club that was able to make instruction available at attractive prices. Sheer adventurism and an interest in serving the defense effort in the event of hostilities appear to have been the primary motivations of most of the members of this group. While the WFA eventually succumbed to the great postwar general aviation bust, it did teach many women to fly. By 1941 it had firmly established itself in ten major U.S. cities.

The person largely responsible for the vitality of WFA was Vita Roth, a former exhibition parachute jumper. Roth was obviously a considerable force in the promotion of women's aviation, but she was also a separatist: "We don't ever expect to be in combat: that's a job for men. It's all bunk to say women are better pilots than men. We aren't. Most women don't have the mechanical talent or physical stamina of average men pilots; but we are qualified for secondary service behind the lines" (Douglas 1991, 5). In 1942, when this statement was made, it may have been a realistic thing for a leader to say. In retrospect, it seems designed to discourage and limit, as if to foretell the collapse of the airwomen's cause just two years later.

To be sure, some attempts at separate service did not involve such a strong implication of inferiority. In 1942, the state of Tennessee formed a program to train female pilots to become military flight instructors. The motivation was similar to that which had underlain the formation of the WASPs: it was expected that the wartime need for combat pilots would be very great, and female flight instructors could release men for battle while maintaining the supply of newly trained pilots. The idea attracted attention nationwide, in part because of the traditional women's role as schoolmarm. To quote then Assistant Secretary of Commerce Robert H. Hinckley about the Tennessee plan, "Women have always been the fundamental background for teaching . . . they can contribute much toward relieving manpower for actual combat fighting" (Leonhirth 1991, 90). Of course, flight instruction had previously not been a female domain; if Phoebe Omlie, as director, had been able to carve out a niche for women in this important area, it would have been a great professional advance.

Omlie was a separatist, but she came to this position out of a desire to achieve the possible, rather than from any philosophical conviction:

> Any woman who can drive a car can learn to drive a plane. It isn't hard and it requires no more skill than it does to operate a car in city traffic. But, while I have a transport pilot's license, I don't fool myself.

> The pilots on our air lines are and will continue to be men. We don't
> hire women chauffeurs and locomotive engineers, no more shall we
> be persuaded to ride behind women air pilots. But that's no reason
> why women should not learn to fly (Callen 1929, 29).

It is unclear whether this statement expresses defeatism or realism. The
latter seems the best interpretation, because other separate female domains established by Omlie — first the Air Marking Program and later
the Tennessee flight instructor program — were not second-class operations at all. They were major parts of the aviation scene.

The Women's Research Flight Instructor School, located in Nashville, duly enrolled ten women in its first class. All graduated in February,
1943, and none had difficulty finding an appropriate position. This demonstration was successful as well as cost-effective. Moreover, it fit well with
a national plan to establish a series of women's flight instructor schools
throughout the nation. Tennessee thus expected federal support for subsequent classes. Congress, however, appropriated no funds for this program; no more women were trained in Nashville or anywhere else. It is
difficult to ascertain why this was so, but timing may have been a factor.
By the end of 1943, it was evident that the combat pilot shortage was not
so great a problem as originally feared. Another possible route to collective
achievement for women was closed off (Leonhirth 1991, 96–97).

The doctrine of separateness thus appears to have been a double-
edged sword. It tended to maintain the most productive areas of aviation
as exclusive male preserves, while emphasizing that women's traditional
responsibilities would compete with an aviation career. It is quite possible
that the airwomen, many of whom believed in separate arenas for women,
were deterred by the limitations implied. On the other hand, the doctrine
did generate separate attention; in the absence of this, perhaps even fewer
opportunities would have come the airwomen's way. Further, it appears
that it might have been possible to manipulate separatism in a very
beneficial way; if areas like flight instruction had become predominantly
female, separate would not have implied unequal. It is by no means clear
that absolute rejection of separatism would have been appropriate; it is
likely, however, that a more sensitive response to its challenges would have
had distinct long-term benefits.

Leadership

The decline of the airwomen cannot be interpreted independent of what happened to other women during this era. In the 1940s, the decade in which the airwomen faded from view, many other women's aspirations went unfulfilled. The airwomen's history might thus be regarded as a single example of the more general case. As noted in Chapter 3, despite the immense disruption and attendant opportunities of World War II, during the 1940s there was a great deal of continuity in the lives of American women. Women worked in unprecedented numbers and moved into the military for the first time, but many events reinforced the dominant position of men. Men faced the dangers of combat and they, as heroes, received the gratitude of the victorious nation. Job preferences and educational benefits were theirs through the GI Bill. This increased gender disparity; for example, by the end of the decade a greater proportion of Americans with college degrees were men than had been the case at the beginning. While women held a more prominent role in the work force, most of their jobs — and this was dictated by economic demand — were in clerical, service, and other relatively low-status roles. These came to be regarded as typical women's work, while more responsible jobs remained largely the preserve of men; a gender-based division of labor persisted even outside the home.

Moreover, many war-related social problems, such as divorce and juvenile delinquency, were in the public mind linked to the breakdown of traditional family values. A chief reason for this breakdown was thought to be women's wartime absence from or lack of devotion to the home. Many people believed that women had been remiss; they were now expected to save civilization by returning to their true responsibilities (Hartmann 1982, 25).

These were very powerful constraints. Moreover, such external forces were coupled with the desire of many women themselves to re-create an idealized prewar reality. Perhaps the failure of the airwomen to advance was simply due to the nature of the times. This implies, however, that nothing could have been done. Humankind is not helpless before such realities. While they certainly make intervention more difficult, they often call forth the very leaders who effect the greatest change.

Many believe that women's leadership was in short supply following World War II. Unfavorable comparisons to the level of political activism following World War I are common: "[The] right to vote came as a direct result of the roles women played in World War I, during which their activity was far shorter and less sacrificial than that of World War II. Women might well have claimed similarly significant benefits after that war — if only they had tried" (Weatherford 1990, 309). The well-documented post–World War II reluctance of women to join the political process in support of equity appears, from today's standpoint, astonishing. Most women's organizations at the time were apolitical. This merely reflected the nature of American life in general, and had nothing to do with gender. However, even those groups that had public policy agendas were often not focused on the well-being of women as a group. The League of Women Voters, for example, specifically denied being a feminist organization and publicly committed itself to the election of the best candidates, regardless of gender. Nor did the women's press take up political challenges of the day: when Congresswoman Mary Norton and Senator Claude Pepper introduced a bill declaring higher pay for men or preferential treatment regarding layoffs to be unfair labor practices, "not even *Independent Woman*, the era's most tireless advocate for women, ran any campaign in support of the legislation" (Weatherford 1990, 309).

Women were, however, recognized as a potential political force by the major parties. After all, they constituted a majority of American voters following the war. Both parties began offering symbolic recognition as early as 1944, when women for the first time presented major addresses at the national conventions (Hartmann 1982, 153). In party politics, as elsewhere, this potential went unrealized, for few tangible concessions were forthcoming. The reasons for this are complex. Certainly sexism was involved; male party leaders did not feel obliged to follow symbolic recognition with any real sharing of power. Indeed, many thought that symbolic recognition was a substitute for power sharing, a way to keep women quiet. Male leaders may also have been aware of a lower rate of turnout on the part of women voters and, as new public opinion polls revealed, a tendency for them to defer to their husbands' political preferences. Women's reluctance to play politics, to use the available potential, was also a contributing factor. Many women, it appears, took the advice of very conservative advisers. "Women should make a study of how not to annoy men in public meetings and discussions" (Pickel 1945, 14).

One way not to annoy men was for women concerned with government and world affairs to retreat to a separate sphere, to take up causes that would ruffle no feathers. Indeed, a premier issue of this kind, and one traditionally feminine, was humanitarianism. Following World War II, a largely failed group from the twenties, the Women's International League for Peace and Freedom, was reborn. According to some people, women detest war and deplore its attendant human suffering to a greater degree than men. They can gratify this peculiar emotional need, goes the thinking, by getting involved in organizations that wring their hands over such matters, articulating goals no one can possibly protest. Just as women were expected to play separate and nonthreatening roles in aviation, they were expected to do so in public affairs. Women in droves joined such groups in preference to those that sought economic benefits and fair treatment for their sisters (Weatherford 1990, 315–316). Humanitarianism of this kind, as Ruth Nichols discovered, went nowhere.

But did the general lack of leadership really mean anything? Had there been more leaders, more energetic participants, would things have been any different or would "the times" have rolled on anyway? The best answer is found in studying the exceptions. Though they served only a minority of women, there were, in the forties and even earlier, vital organizations that overtly campaigned for professional advance and gender equity. The most prominent, perhaps, were the National Women Lawyers Association, the American Medical Women's Association, the American Association of University Women, and the National Federation of Business and Professional Women's Clubs. Women in working-class occupations were represented by the National Women's Trade Union League. These organizations functioned like most interest groups in traditional American politics, carrying their story to various government officials, seeking a sympathetic ear, and hoping to exert subtle influence.

These organizations were successful in a limited way, consistent with their diminutive size. Beginning in 1942, coalitions of these groups systematically attempted to influence the appointment of women to government positions. Their efforts resulted in a 1944 White House Conference on Women in Policy Making and spawned a fairly extensive network of monitors of various agencies. They are credited with "a modest expansion of opportunities for women in the federal government" (Hartmann 1982, 149). Similar efforts to promote female electoral candidates also enjoyed modest success: the total number of women serving in all the state legislatures rose from 144 in 1941, to 249 in 1950. Women in Congress were increasingly elected on their own merits, rather than, as was often the case in the twenties and thirties, being chosen to replace their deceased spouses. Also in contrast to the past record, virtually all women in the Congress sponsored legislation advancing gender equity in one way or another.

To be sure, this record is one of small positive steps, rather than massive and celebrated accomplishment. Perhaps more important than immediate success is the question of a lasting legacy. In the forties, there were vigorous, public-policy-oriented, gender-sensitive organizations of lawyers, doctors, and businesswomen; there was no counterpart that embraced the airwomen. In current times, lawyers, doctors, and business-women from this period are recognized in biographical dictionaries of notable women; airwomen, by and large, are not. Arguably, this is not a coincidence.

Is it too much to think that pilots, instead of duplicating the inertia of most women in the forties, might have joined their more active sisters in this search for gender equity? Indeed, there was an established organi-zation, the Ninety-Nines, and there were a number of potential leaders of some accomplishment and visibility. The potential of these sources of leadership was not, however, fully exploited. Why this was so ultimately involves the personalities of key individuals — Blanche Noyes, for exam-ple, as president of the Ninety-Nines, and Jacqueline Cochran. As Chap-ter 3 clearly indicates, Cochran is by far the most interesting subject for analysis and, unlike Noyes, considerable written material by or about her survives.

Cochran was an ambitious and aggressive person; these qualities had a great deal to do with her success. Desirable as ambition and aggression are, however, they were not, in this case, well-harnessed to broader goals of gender equity. As Chapter 3 argues, Cochran was into aviation for her own gain. While her demonstration of women's aerial competence was indeed marvelous, she did not make any systematic contribution to women's professional advancement.

There appear to be several reasons for this. The most basic is perhaps Cochran's strong need to control and manipulate others. Such a need is not necessarily bad; indeed, it is probably part of the motivation of strong leaders everywhere. Cochran's laudable initiative in organizing the WASPs certainly derived, in part, from a desire to sell her vision of the future to others. Difficulties arose when Cochran had to confront the realities of a complex world. Successful leaders quickly discover that success requires something in addition to manipulativeness or the application of control. Most obvious is the necessity of compromise. In the modern world and especially in democratic societies, few results can be produced without compromise. By definition, compromise involves partial surrender of control. When in a leader's mind, control is more important than end results, trouble is on the horizon. If, in the long term, people are to respond to leadership, if they are to submit to control, they must receive something in exchange. In modern terminology, this is called empowerment of followers. Leaders who are not sensitive to what followers want, who are not prepared to incorporate followers' needs into their visions may eventually find that they have no one to lead.

Many have faulted Cochran for her failure of leadership in the twin issues of WASP militarization and deactivation. As noted earlier, the WASPs remained civilians. Cochran, like many others, was a strong proponent of militarization. There were many important considerations. Among them, for the pilots at least, were pay and benefits; for the Army brass, elimination of an organizational anomaly. More generally, clear to anyone with vision, there were questions of prestige for and formal national acknowledgment of female aviators.

One suggestion for resolving the WASPs' peculiar status was to incorporate them into the highly successful Women's Army Corps (WAC). Though in some respects attractive, such a solution would have made it necessary to forge a connection between the WAC and the Army Air Forces, the latter being the unit the WASPs served. Reporting lines would have changed such that Cochran would have reported to Oveta Culp Hobby, Director of the WAC, rather than to General Arnold as she did as civilian Director of Women Pilots. Cochran's opposition to this plan was understandable, but the bitterness and intensity of that opposition were surprising.

Cochran justified her position on the basis of Hobby's character and Hobby's lack of familiarity with aviation. Cochran admitted that Hobby was the "woman I loved to hate" (Cochran and Brinley 1987, 198), but provided no substantive rationale. Hobby allegedly knew little about aviation and certainly did not share Cochran's notion that female pilots were fundamentally different from other military women. As Cochran put it, "If she didn't understand airplanes or the kind of women who flew them, then she had no business being their commanding officer" (Cochran and Brinley 1987, 207). This notion is suspect on a number of fronts. First, Cochran presented no evidence of the unique character of female pilots. Second, it is simply impossible for all officers in a complex line of command to have personal familiarity with every function performed by constituent units. Whether Hobby understood aviation had nothing to do with her competency to command the WASPs; an individual in this kind of high-level leadership position needs to understand organizations, human nature, and politics.

Even if Cochran's assessment had been valid, hating someone who did not share her views was counterproductive. It makes more sense to attempt to persuade rather than to hate; it may be necessary to preserve the possibility of a working relationship should compromise become necessary. One suspects that Cochran hated Hobby out of a fear of partial loss of control. In order to protect her emotional need, she cut the WASPs off from one possible solution to the militarization dilemma.

Cochran was similarly intransigent with regard to the function of the WASPs. This organization not only deployed established female pilots to serve in ferrying or other functions, but it also trained women to fly military airplanes. It was to be a self-regenerating and permanent unit. While the comprehensiveness of the organization made sense early in the war when battle casualties and the attendant demand for pilots were expected to be high, it most assuredly did not by 1944. Pilot training of all sorts, including that for male combat types, was being scaled back. Indeed, male instructors left unemployed in the wake of a combat training cutback generated much of the anti-WASP political pressure that year. In early Congressional discussions of the military status and long-term future of the WASPs, there was considerable support for incorporating those WASPs already in service into the Army (Douglas 1991, 55). Although this represented an opportunity for a significant gain, Cochran refused to consider it. She insisted that the training function remain firmly and completely under her control, and linked this to a demand for military status separate from the WAC. Why would she not compromise on training when it was so obviously an unrealistic demand? Again, a plausible answer is that she had lost sight of the goal of providing military benefits for her charges, of the need to advance the long-term professional prospects of female pilots. Instead she focused on the necessity of maintaining control on her own terms.

Evidence from other episodes in Cochran's life underscores her need for control. Her autobiographies make clear that she regularly manipulated her nominal friends and associates and tried to intimidate those not immediately useful to her. This, of course, is not unusual, nor is it in all cases undesirable. However, Cochran's writings show that she fiendishly enjoyed such encounters. At about thirteen years of age, Cochran took the first big step in her escape from child labor in the rural South. By falsely claiming credentials ("I added and subtracted information at will, as it suited me"), she managed to land a job in a beauty parlor in Columbus, Georgia. Once established, she demanded and received higher pay by blackmailing her employer; she threatened to inform the state child welfare board of her underage status (Cochran and Brinley 1987, 39). Cochran took pleasure in succeeding through guile and manipulation; she states, with obvious pleasure, "I had her caught."

The same personality trait apparently persisted throughout Cochran's lifetime. When the WASPs were deactivated, Cochran showed no further interest in female pilots. In neither of her autobiographies does she express any concern for the plight of the WASPs, either in the immediate aftermath of the war or in terms of the future of women in aviation. In fact, she confirmed her indifference by leaving the country as soon as possible. Within a few days of the WASPs' demise, she secured orders from General Arnold that allowed her to visit military theaters throughout the world as the war wound down. The nominal cover was a position as correspondent with *Liberty* magazine. Again, this involved a manipulative scam; Cochran bragged, "I don't know how to write, but I'll get to Asia this way" (Cochran and Brinley 1987, 215).

Cochran had no particular reason for going, except to tour. However, she enjoyed using Arnold's patronage to put down or extract favors from various American generals in the field (Cochran 1954, 167). In seeking permission for correspondents (herself included) to remain in certain areas of China, she had to deal with a General Stratemeyer. He was reluctant to grant permission. Cochran's response: "He didn't know me. I had already drafted a cable to my friend, Secretary of War Robert Patterson" (Cochran and Brinley 1987, 227). To follow up this victory, Cochran further humiliated Stratemeyer by flying off in grand style: "He couldn't believe that General Kenney, air chief in the Far East, had put a plane at my disposal."

In 1959, Cochran took her own plane to the Soviet Union. She was not permitted to overfly restricted military areas or nuclear facilities and was required to take on a local navigator to ensure that she did not. Cochran resented this intrusion on her control and decided to punish him; she spends three pages of her autobiography describing how she ordered this "boy" around and enjoyed his growing terror as she gratuitously allowed ice to build up on the plane during a storm (Cochran and Brinley 1987, 293–296).

The examples could be multiplied, but the point is already clear: Cochran was a control junkie. When the WASP operation no longer gratified that impulse, she quickly went someplace else to get a fix. This general concern with manipulativeness and control is probably related to Cochran's orientation toward women as a group. Cochran clearly enjoyed being a protector and benefactor to women whom she perceived to be of lower status. As noted earlier, she financed the first postwar AWTAR races for women but would not deign to participate in any of the events sponsored by the Ninety-Nines.

Cochran's protection and chaperoning of the WASPs is well known. Her regulation of the personal lives of the recruits and concern with sexual impropriety earned Avenger Field the nickname, "Cochran's convent" (Holm 1982, 144). This tendency sometimes led to faintly ridiculous events, such as her attempt to shield her American ATA girls from the probing eyes of British medical examiners. All ATA personnel were required to undergo physical examinations that necessitated disrobing. Cochran protested, with the help of well-placed contacts, until her charges were granted exceptions. The ATA girls themselves found Cochran to be prudish and unduly restrictive (Cochran and Brinley 1987, 190).

Cochran undertook another Lady Bountiful effort in the early sixties. Jerri Cobb and a few other women sought acceptance into the U.S. Astronaut Program, setting off another protracted period of controversy about women's abilities and proper roles. At first, Cochran seemed to be a great ally, a powerful advocate of women in space. She had once stated that she would "have given her right eye to be an astronaut" herself (Cochran and Brinley 1987, 259). Moreover, Cochran helped to pressure a reluctant NASA by paying the considerable costs of both the physical and mental astronaut test programs on behalf of the female applicants. This certainly seems laudable, but a strange change occurred. Ultimately, Cochran argued against women's involvement in the space program, most emphatically against their serving as astronauts. In her view, a separate women's unit might perform some functions in the space program, but such a unit would be too expensive. Moreover, she argued that women were not sufficiently committed because of the competing demands of marriage and family. Astonishingly, she wrote to Cobb, "Women for one reason or another have always come into each phase of aviation a little behind their brothers. They should, I believe, accept this delay and not get into the hair of the public authority about it" (Douglas 1992, 81).

Why this turnabout? It is possible that Cochran realized that successful female astronauts would be regarded as dramatic achievers in their own right. Such figures would not for long be gratefully dependent on Cochran's approval or bounty. They would not be Cochran's wards, but Cochran's competitors. Such women, in Cochran's view, were dangerous.

This behavior is consistent with Congressional testimony Cochran gave in 1974 concerning the wisdom of admitting women to the U.S. military service academies. Cochran was opposed, especially with regard to the Air Force Academy. Despite her own earlier desire to serve in combat (Cochran 1954, 134), she argued that women should never face actual fighting; since the academies train for combat, women must be excluded (Holm 1982, 308). This is a fallacious argument, because many academy graduates are trained for roles that are well removed from any combat sphere. Perhaps it is more to the point that closing the doors of the academies lowers the probability of women developing distinguished military careers, careers that could rival Cochran's.

That is to say, women who might achieve the same status presented a problem for Cochran. We have already seen that she generally ignored other airwomen, even to the extent of failing to acknowledge Louise Thaden, the only other woman to win the prestigious Bendix race. As Chapter 3 demonstrates, this general disregard continued into the postwar years. When it was impossible to ignore other women's aerial achievements, Cochran reacted with anger and hostility. Cochran had been involved in the ferrying of a Hudson bomber to Europe in 1942, a feat described in Chapter 1. She was very gratified by this accomplishment, but it appears that her pride was not in the fact that it symbolically opened the door to new achievements by women, but rather in the fact that she was the only woman to do this. Indeed, the section in her autobiography describing the event is entitled, "Not Everyone Flies Bombers to Britain" (Cochran and Brinley 1987, 167).

WASPs Nancy Love and Betty Gillies also attempted a bomber run to Britain as part of a plan to move the most experienced ferry pilots into the most demanding assignments regardless of gender. Though the event was scrubbed by General Arnold, the very attempt is reported to have made Cochran furious (Verges 1991, 140–141). When Love and Gillies began checking out powerful P-47s and P-51s in February 1943, this too caused great upset to Cochran (Granger 1991, 66). Cochran was ostensibly concerned about overall esprit de corps when, in reaction to events like this, she said, "Individual cometlike achievements should be avoided, graduation into important new assignments should not be by exceptional individuals, but by groups" (Verges 1991, 141). She also expressed concern about bad publicity should Gillies or Love have a mishap in advanced aircraft. These thoughts, however, have a hollow ring. They are contrary to what Cochran herself did as an active airwoman. Moreover, Cochran immediately afterward vigorously argued for the assignment of WASPs to B-17s — albeit in a training command role that she would plan and organize (Verges 1991, 142). It is much more likely that Cochran wanted to be the number one woman and couldn't stand the possibility that she would have to share the female spotlight.

Cochran also had very strong feelings about women's appearance and general demeanor. Conventional attractiveness and man-pleasing femininity were important both in terms of her own person and in terms of her expectations for others. Cochran had been involved with beauty parlors, both as an employee and eventually as an owner of several successful businesses. In addition, she established one of the largest and most successful cosmetics firms in the United States. This work was very important to her — indeed, she almost appeared to believe that the contribution of her cosmetics to womankind was considerably more than the contribution of her aviation accomplishments: "We offered our customers a lot. . . . It's a woman's duty to be as presentable as her circumstances of time and purse permit" (Cochran and Brinley 1987, 119).

Cochran emphasized her own femininity, spent a good deal of time looking after her appearance, and in general affected a demeanor that was pleasing to men. She at times carried this notion to extremes; she wrote that immediately following the conclusion of an air race, "I refused to get out of the plane until I had removed my flying suit and used my cosmetic kit. This was feminine" (Cochran and Brinley 1987, 20). Cochran and Brinley cite with apparent approval the assessment of General Fred Ascani, postwar test pilot and one of Cochran's patrons as she attempted her various record flights in military jets: "She could be very soft, very feminine. I danced with her once in Spain and she was as female as any other woman I've put my arms around. . . . She was a man's woman" (Cochran and Brinley 1987, 21). Cochran's concern extended to her charges as well. She wanted the WASPs to conform to her stereotype and went to considerable lengths to arrange and enforce this. Perhaps the best-known event demonstrating this was the selection of official WASP uniforms. They were ultimately produced by Nieman-Marcus on a custom-fit basis (Verges 1991, 167). This concession to Cochran's vanity turned out to be significant; the uniforms were very expensive and many regarded them as an unjustified extravagance during general wartime hardship. They generated resentment and dismay throughout the military and provided grist for the WASPs' opponents.

Cochran's concern with appearance is evident in events that occurred at other times in her life. The most serious of these occurred well after World War II. The Air Force had separated itself from the Army in

1947 to become a distinct unit of the armed forces. The new Air Force planned to recruit women on a regular basis, though not as pilots. WAF, Women in the Air Force (not to be confused with WAFS, the Women's Auxiliary Ferrying Squadron of 1942), appeared in a number of areas. Though the Air Force was anxious to integrate WAFs into the regular military structure in order to facilitate clear lines of command, this had not been definitively accomplished by 1950. For example, there was a separate director of the WAFs. In a number of other ways, including relatively low numbers of recruits, the WAF operation had not lived up to initial expectations. Some sort of review and redirection was clearly necessary. Chief of Staff General Hoyt Vandenberg called on Jacqueline Cochran as an outside consultant.

Cochran's report did not address job performance or productivity; instead, Cochran criticized poor appearance and lack of grooming, which she asserted was rampant among the WAFs. When she visited WAFs on site, she was offended by, for example, the "hormonally maladjusted" women she encountered, as well as those who were cross-eyed. These she labelled as misfits. Cochran's proposed solutions involved screening on the basis of appearance and the establishment of a separate WAF organization presided over by a woman in a very high position, perhaps Assistant Secretary of the Air Force.

Vandenberg attempted to implement some of these suggestions, but even in 1950 they were widely regarded as outrageous. The general was discreetly reminded by subordinates that there were no Air Force regulations that permitted dismissal on the basis of appearance. The report may have caused a temporary retreat from the goal of gender integration in the Air Force, but no structures resembling those preferred by Cochran ever came to pass (Holm 1982, 137–147; Douglas 1991, 70–71).

Clearly, Cochran's views had changed little since the days of the WASPs. She believed that women in the military had to be attractively feminine. Just as the Women's Auxiliary Ferrying Squadron could not be integrated into the Air Transport Command in 1942, so must the WAFs be held separate from the line commands of the Air Force in 1951. Just as the Director of Women Pilots must run the WASPs from a special position at the right hand of the highest ranking Army Air Forces general, so must the WAFs answer to a uniquely connected woman not subject to regular officers. The only difference was that by 1951, Cochran's view of womanhood was more widely viewed as anachronistic than it had been in 1942.

It is clear that Jacqueline Cochran could not meet the need of the airwomen for postwar leadership. Though she was by far the most likely candidate to head a group of professionally ambitious female pilots, she was unequal to the task. Cochran's strong need for control and domination led her to avoid compromise and eschew possibly beneficial alliances when the WASPs' future became an issue. It appears to have driven her to indefensible positions. Moreover, this basic personality feature gave Cochran a particular orientation toward women: for those of lower status, she would be a protector or benefactor, but successful women, even if the product of her personal largesse, appeared to represent an unacceptable threat.

In addition, Cochran had a strong attachment to traditional femininity. Her own role as an airwoman, of course, was immensely out of accord with traditional expectations for women's behavior, as was her confidence that women could fly military airplanes. But her view that this radically nontraditional activity should be performed only by those who held to old-fashioned symbols was anomalous. It was dangerous as well, for it caused Cochran to reject many legitimate women's interests and to neglect potential allies. Moreover, these symbols of traditional femininity caused resentment among those who, between 1942 and 1944, had a war to win.

What function did this emphasis on physical attractiveness and traditional femininity play in the larger areas of Cochran's personality? A remark Cochran made to her WASPs provides a clue: "When a man wants to put your parachute in the airplane and take it out, let him. That's what men are for — to be nice to us. If you are going to run around trying to act like men, they are going to treat us like men. If we act like ladies, we'll be treated that way" (Cochran and Brinley 1987, 214). Cochran felt that in order to fly, she needed the help and support of men. To get this support, she felt it was necessary to do all the traditional man-pleasing things. In short, Cochran used her sex to manipulate men. Given a personality that thrived on manipulation in general, it must have been relatively easy.

Conclusion

The saga of the airwomen might have ended differently, and their fame might have lasted longer. Why these things did not come to pass is indeed a complex question.

The airwomen showed a remarkable tendency to retire at an early age. Many airwomen faced great obstacles not of their own making, chief among them traditional views about family and domestic life. On the other hand, many of these obstacles did not seem inherently overpowering. Some airwomen, like Louise Thaden, returned to active piloting time and again, at least temporarily overcoming restraints. Others, like Betty Gillies, demonstrated that it was quite possible to meld family and a flying career. One can hardly fault the airwomen for a general lack of ambition, but it seems that just under the surface there was potential for greater performance. What might have induced Ruth Law and Neta Snook Southern to challenge limits imposed by their families? What might have prevented Elinor Smith's fifty-year indifference? A good candidate for the missing ingredient is leadership.

The airwomen knew that they were being asked to play a special and potentially demeaning role in the maturation of aviation. They were supposed to demonstrate that flying was safe and piloting easy. By and large, airwomen did not object to this assignment; indeed, they embraced the professional opportunities it created. It is far too much to expect that they would have proudly disdained these jobs in favor of withdrawal from aviation altogether — and indeed, it is not clear how that course of action would have better served the cause of women in aviation. It does appear, however, that the matter was not kept in proper perspective. The more insensitive uses of female pilots in this way were generally orchestrated by men who were indifferent to the long-term well-being of the airwomen. However, few female pilots were sounding the alarm. Again, while the overall behavior of the airwomen can hardly be faulted, is is unfortunate that a leader with vision did not emerge.

The history of the doctrine of separateness runs in direct parallel. The establishment of separate races and separate record categories for women generated a great deal of favorable attention. Yet the representation of women as a group with distinctly separate domains, particularly domestic ones, by implication excluded women from the more prestigious aviation pursuits. It is regrettable that the implications of this dichotomy were not more clearly seen. Phoebe Omlie was unusual in believing that if women must develop separate aviation spheres, they should be important ones that carried no implication of inferiority. Her establishment of the Air Marking Program was a step in the right direction, as was her Tennessee flight instructors program. Had the latter been continued, women's separate sphere might have been impressive indeed. Unfortunately, this sort of leadership vision was in short supply.

A direct consideration of leadership among the airwomen reveals a similar sad picture. Jacqueline Cochran was well positioned in terms of ability, reputation, connections. Moreover, she was at the peak of her powers at the time leadership was most urgently needed. Although immensely successful and quite anxious to have an impact on events, Cochran had some very basic personal characteristics that prevented her from accepting responsibility for the airwomen's future. She was oriented to a different time and a different clientele.

The times were difficult and dangerous, but at least around the edges some gender-related traditions were breaking down. The opportunity to define a new professional future for the airwomen seemed within grasp. The airwomen were present in some numbers and might have responded, but they were not called.

Of course, had appropriate leaders emerged, sexism would not thereby have been eliminated, nor would male hostility to the airwomen have dissipated. Injustice and disappointment would certainly have continued. Nonetheless, projection of a vision, active lobbying, and organizational support might have stimulated a few to professional advance even in the face of bigotry. More important, perhaps, these activities might have secured a place for the airwomen in the long-term collective consciousness of the culture. As it was, however, within days of December 20, 1944, the airwomen slipped into inactivity; within months they fell into an obscurity that has persisted nearly to the present day.

✿ 6 ✿

Belated Legacy:
Promise Reborn

In recent decades, professional female aviators have, in modest numbers, come into their own. As the scant legacy left by the airwomen would suggest, today's airline pilots owe little to their predecessors. They have developed independently, without any direct help from their professional ancestors, and often with little significant knowledge of them. It might not seem too farfetched to consider the airwomen historically irrelevant; the future that they might have shaped ultimately developed without them.

Indeed, their lack of influence underlies the primary message of this book. This lack of impact should be marked well and deeply understood so that something similar may not happen again. While today's female pilots demonstrate how our society has progressed, it is crucial to recognize that the culture would be even more advanced today had the airwomen not been shunted aside. The story of the airwomen provides a wonderful object lesson, albeit a negative one.

The world, however, is full of ironic twists. As noted at the close of Chapter 4, there are some hints that the airwomen's failure to leave a long-term legacy might not be fixed for all time. This chapter examines trends that suggest the airwomen may someday emerge from the mist. Thus, this book offers another message, and a more hopeful one.

In order to discern these developments, this chapter focuses on trends in the popular literature and communications media, in the hopes and dreams of educators and activists, rather than on historical analysis. Many commentators, and just ordinary people, are discovering the airwomen as if for the first time. What we see is celebration, a newfound desire to venerate and honor the airwomen. Nothing of this sort has appeared since before 1944.

The current revival may not be a stunning revolution, but it is complex, so much so that it is a little hard to grasp all of its dimensions. The remaining airwomen, even though some of them are explicitly contributing to the present ferment, are baffled by it. It is visible in a host of new books and other publications, in memorials and museums, and in aggressive promotion schemes.

Communications Media

Very little about the airwomen was published in the forties and fifties. In the sixties, however, popular writers on a vast variety of topics developed a new genre: very brief biographical sketches of noteworthy people, either famous or unusual. This may have been in response to the conditioning of the American public to rapid-fire, low-depth treatment occasioned by nearly two decades of television. In any event, unique personalities were sought out as subjects for exciting, factual reports. Although there was no great rush to make celebrities of them, the airwomen joined their counterparts from many walks of life as vague curiosities. It was the development of this literary form that occasioned the return of the airwomen to the printed page. Aviation magazines featured short vignettes, often in an informal series, about several airwomen (Buffington 1966, 1967, 1968a, 1968b, 1969; Chun 1969). Some writers saw the opportunity to string several such accounts into books, and a lengthy compilation on airwomen appeared as early as 1962. May's work of that date is not particularly notable because it is totally devoid of interpretation; it simply and laboriously presents facts about various airwomen. As the first of a new literature on the airwomen, it is not an inspiring example, but it was a beginning.

Magazine articles on particular airwomen continued to appear in small quantities (Valerioti 1984) over the years and, with considerable improvement in quality, even moved into high circulation and upscale publications (Gwynn-Jones 1984; Pateman 1985). They remain to the present day (Chase 1990). Books that devote one chapter to each individual likewise established themselves as a permanent form. Some addressed female pilots alone (Jablonski 1968; Hodgman and Djabbroff 1981; Lomax 1987), while general compilations about early aviators also featured notable airwomen (Vecsey and Dade 1979). Such titles continue to appear (Holden 1991). They have achieved varying degrees of success. Most suffer all the disadvantages of the type; lack of integration and obscurity of purpose explain their rather limited impact.

In the late seventies and early eighties, however, more sustained treatments of the airwomen began to appear in several forms. Autobiography was one of the most prominent. Elinor Smith had been invisible for well over forty years when she broke her silence in 1981 with a full-length book detailing most of her flying career. She does not reveal why 1981, rather than 1971 or 1961, was the publication date. Smith's work did not stand alone; a decade later Edna Gardner Whyte published her memoirs (Whyte and Cooper 1991). Whyte's delayed schedule is more understandable because she was active in aviation for a much longer time. Whyte decided to set her life experiences on paper when her flying career began to wind down. Her motivation is also somewhat more clear than Smith's. Whyte has a message for contemporary society; she is concerned about gender equity and, unlike Smith, is quite explicit in urging today's women to greater achievements.

Three of the youngest airwomen have prepared accounts of the WASPs that, though intended to be generally expository, have a large measure of autobiographical focus. Scharr (1986, 1988) wrote two large volumes from copious notes taken during her years of service. Her books purport to be history and to reveal previously unknown material; they are however, mostly of personal significance. Scharr claims to have come from a relatively humble background, in sharp contrast to the wealth and social position of her colleagues. Scharr thus did not get along well with the WAFS establishment. She felt she was the object of snobbery and prejudice, and her potentials were not permitted their fullest development. Her books are rather bitter, dedicated to frustration and belated assignment of blame. Perhaps this is unfortunate, but if bitterness was the stimulus to write, it may have served a very positive function.

Granger (1991) regards her thick volume as a definitive factual treatise. It contains a great mass of systematic data, much in appendix form. Comprehensive lists of WASP duty assignments and base locations are some examples of the information she provides. Badly organized and without central themes, the book does indulge in interpretation, or perhaps one should say assertion. The personal character of many of the players in the WASP drama was Granger's real interest. Vindication of personalities seems to be the purpose. In most respects, this misses the point; whether individual people were nice or nasty is of little relevance to the meaning or impact of the WASPs. This is also a badly flawed effort. But flawed or not, we have at last some examples of first-person contributions to a long-neglected story.

Cole's (1992) shorter book is rather good natured when compared to Scharr's or Granger's. The book reads like an oral history, telling stories of the WASPs for their own sake. It is as though someone had asked Cole to tell us how it was back then, and Cole, microphone in hand, had obligingly responded. Reading Cole's work is like listening to the fascinating tales of a grandparent.

In literature, a revival may be signified by the reissue of titles from a previous day. This is the case, to a limited degree, with the airwomen. Beryl Markham, a British Kenyan, was technically not a U.S. airwoman, but had important American dimensions to her career. Markham had been a horse trainer in Kenya, a member of the white establishment. She was one of the first pilots in Kenya and certainly the first female pilot in Africa. She become a well-respected bush pilot and eventually gained the friendship and patronage of the colonial elite and prominent Britons, including members of the royal family.

As noted in Chapter 2, Markham flew solo across the Atlantic westbound in 1937, a notable feat. Remaining in the United States, she wrote her autobiography, which was first published in 1942. There is some controversy over this book because it is extremely well written, considered by many to be a literary masterpiece; since there is little other evidence of Markham's writing ability, many suspect it was ghostwritten. Indeed, there are certain passages that do not seem to have been composed by a pilot — for example, one that suggests that a search for a missing party would have to be flown at a low level, rather than from the vantage of a high altitude. It is impossible to know whether Markham embellished in order to make the book more dramatic or whether she commissioned someone to write it. For our purposes, this is not of primary importance. More to the point is the fact that an American entrepreneur chose to reissue *West With the Night* in 1983. An airwoman's story was available to the American public in comprehensive form for the first time in many years. The publisher reportedly became very rich. Many contemporary female airline pilots have read this book.

The pattern for biographies is similar. Such publications have been appearing at an increased rate in the past decade (Backus 1982; Shore 1987; Morrissey and Osborne 1988; Lovell 1989, Rich 1989). Earhart biographies are a special case; they have, it seems, always been popular, even when nothing else on airwomen was being published (Morrissey 1963; Briand 1966).

Although such efforts are not, strictly speaking, biography, it is interesting to note books that focus on Earhart's disappearance over the Pacific. Stories of a spy mission, capture by the Japanese, a secret return to the United States, or other unsolved elements of the Earhart trip (usually involving a sinister coverup) have been around since 1937. Book-length treatments of the "mystery" have appeared at various times over the past twenty-five years (Goerner 1966; Klass 1970). The release of titles with this theme follows the same pattern as traditional biography; in the past few years, many authors have again attempted to solve the mystery for us (Brennen 1988; Devine and Daley 1987; Donahue 1987; Loomis and Ethell 1985; Myers 1985). Cynics might question the quality of research that produced some of these volumes, and others might charge them with sensationalism, but it is certainly true, if the demand for such literature is any guide, that interest in this particular airwoman has lately grown by leaps and bounds.

This increase is part of a general resurgence of interest, not solely an Earhart-specific phenomenon. It is paralleled by the emergence of biographies on three other airwomen, whereas no book-length treatment of any of them had ever appeared before. Grover Ted Tate's short work on Pancho Barnes, published in 1984, is a very personal document based on the friendship between the author and the subject. During his military years, Tate was apparently a regular at Barnes's Happy Bottom Riding Club. The book highlights the most unconventional features of Barnes's personality and celebrates her various conflicts with prominent citizens as well as her challenges to established social mores. The main sources are interviews with Barnes completed in the last years of her life.

Donna Veca and Skip Mazzio (1987) prepared an unusual volume about Bobbi Trout. It is mostly a collection of photographs, newspaper articles, and other memorabilia; frankly, it resembles the contents of a box long stored in an attic rather than a serious exposition. Explanatory or interpretive text is minimal. An effort to record Trout's comments on videotape appears to be a parallel project of publisher Carol Osborne. Osborne has befriended Trout and thus gained access to her papers and her memories. If the book is part of an attempt to put those papers and memories into some sort of orderly array for posterity, it is only a very modest success.

Also in 1987, Mary Ann Bucknum Brinley published a book billed as an autobiography of Jacqueline Cochran; the book is alleged to be coauthored by Cochran, though Cochran had been dead for nearly a decade at the time of publication. Perhaps the extensive use of quotes from Cochran justifies this nomenclature. The book depends heavily on comments, some from the printed record and some specifically solicited for this volume, of various people who knew Cochran. The book is based on a fabulous wealth of primary sources, but it is rather difficult to identify a basic theme.

None of these efforts can be regarded as a fine example of biography, but that in itself is instructive. These books are uncritical celebrations. Each reads as if it were written by an enthused contemporary greatly stimulated by a powerful personality. The authors seem a little unsure about what to make of their subject, but they know she is someone remarkable. That is, the authors seem a bit like disciples; they are certainly not thoughtful commentators who write with the benefit of long observation and hindsight. Instead of reflecting on careers long ago completed, these books appear to be introducing rising stars. This incongruity is perhaps best explained by the fact that the airwomen are only now becoming rising stars — naive enthusiasm perhaps is necessary and appropriate, since it was largely absent in the intervening four decades.

In 1979, Sally Van Wegenen Keil produced the first book-length exposition about the WASPs. Her effort was apparently inspired by an aunt who had flown B-17s in the war. As befits a first work, it is descriptive rather than interpretive. It also emphasizes the human-interest dimension; in a given situation, specific quotes of dubious authenticity are attributed to particular people, and individual emotional reactions to events are described. Quite possibly these represent journalistic license, but the effect seems quite appropriate. In 1991 Marianne Verges prepared a very similar book. It too presents the WASPs in very readable style, flawed only slightly by factual errors. It is noteworthy that interest in the WASPs has risen to a degree that justified another general book a decade after the first.

One of the most interesting recent developments has been the Smithsonian Institution United States *Women in Aviation* series (Oakes 1978; Brooks-Pazmany 1983; Oakes 1985; Douglas 1991). Each volume addresses a defined time period, with total coverage from the dawn of flight to 1985. They are carefully researched, drawing on previously unexploited materials at the National Air and Space Museum. These titles represent an admirable effort at sustained presentation of the female fliers of the United States. Douglas advances a thoughtful, interpretive message by placing her material in a broader context of women's history, but the general thrust of the series is to report clearly and concisely a host of factual data. It is indeed significant that the nation's central archive decided to issue such a clear endorsement of the importance of the airwomen.

During the past decade, airwomen have even inspired best-selling novelists. Janet Dailey's *Silver Wings Santiago Blue* (1984) weaves themes of self-development, emotion, and courage with sheer adventurism in a story about the WASPs. Resembling a fictionalized version of Keil's work, it presents the WASPs as a dedicated and thoroughly admirable group — and it underscores that point with an epilogue on the 1977 establishment of veterans' benefits for the WASPs. Judith Krantz explores an American family, a mother and two daughters, during the World War II years in *Till We Meet Again* (1988). Built around high adventure, the book struggles a bit for aeronautical authenticity, though the author did consult with aviation experts, including surviving airwomen. One of the daughters, Freddy, becomes a crack pilot in the late thirties. Her quest for adventure leads her to Britain, the Air Transport Auxiliary, and various exploits. Freddy's role is the most developed, and Krantz includes a brief prefatory acknowledgment of the many ATA pilots. One of the six World War II female heroes in Marge Piercy's immensely popular *Gone to Soldiers* (1987) is a WASP pilot. This novel explores ten characters who experienced World War II, each in a different way. Bernice finds in the Womens' Airforce Service Pilots a way to escape an authoritarian father. The war changes her greatly, a theme common to all the characters in this very long book. Although Bernice is by no means heroic and certainly not always admirable, it is interesting that the role of airwoman would be chosen by a major author to project colorful themes of war and human experience.

These are not the only contemporary works in which the airwomen receive attention. For example, Anne Noggle's (1990) large-format book featuring photographs of individual WASPs taken in 1988 has received a fair amount of attention, including some in such an unlikely place as the *Chronicle of Higher Education* (Nov. 27, 1991). Airwomen have even invaded the little magazines that insurance companies publish for their clients (Baty 1990).

Most mention of the airwomen in children's literature features Earhart. An occasional young people's book about her emerged in the fifties, sixties, and early seventies. But in a pattern that has by now become familiar, there has been a great increase beginning in the mid-eighties. At least nine Earhart titles emerged between 1984 and 1989 (Niekamp 1990).

Finally, there has been a recent incursion into video. In October 1988, CBS aired a made-for-television movie entitled *Pancho Barnes*. This drama drew on many events in the life of that colorful airwoman. Barnes is portrayed not only as an outstanding aerial performer, but as an admirable example of courage and individual development as well. Her challenge to a domineering husband rates an audience cheer, as does her resistance to Howard Hughes. In fact, Barnes was a movie stunt pilot in some of Hughes's productions and did differ with him on many occasions. The producers, however, took considerable liberty with this fact and rather romanticized not only Barnes but the era in general. Of course, this is not atypical, for the purpose of movies of this sort is entertainment and not historical accuracy. The important point is that a major network would invest in a three-hour, prime-time production that painted a very favorable picture not only of Barnes, but also of many other airwomen of her time. The network's commitment was underscored by their assignment of the starring role to Valerie Bertinelli, a very well-known television actress of the eighties.

A commercially available videotape, *Women of Courage* (Magid 1990), has attracted considerable attention. It is a documentary on the WASPs, featuring footage from World War II and commentary by many of the WASPs who remain active today. Apparently, a national public television airing of the tape has been sought.

As far as the popular media are concerned — both printed and electronic — the airwomen have returned. Of course, the significance of this new output is, at this early date, not entirely clear. Certainly, it is diverse; there seems to have been tremendous variety in author motivation and target audiences. It is clear, however, that this new attention to the airwomen is not the result of some idiosyncracies on the part of a few authors; paralleling the general media, there has also been renewed attention to the airwomen in a number of other areas.

Service of the Aviation Community

In the 1990s, the number of women who make their living in aviation-related occupations is growing. They are exceptionally able and ambitious people, and have formed appropriate professional organizations to better represent their interests — a step that the airwomen failed to take in 1944. The accoutrements of such organizations, including house publications, are clearly in evidence.

Of particular note, however, is the emergence of a general publication for any woman who might be involved in aviation. In 1988, a former National Transportation Safety Board accident investigator started *Women in Aviation*. This twelve- to sixteen-page magazine features articles on, among other topics, notable achievements and professional advancements of female professionals, travel, educational opportunities, government regulations, and flying safety. Book reviews and commentary also grace its pages. Publisher Amy Carmien, a pilot since her teens and holder of an A and P rating, has undertaken many aviation roles throughout her life. Most of the time, she has had few female colleagues. The pervasive sense of isolation led Carmien to reach out to other invisible women in aviation fields.

Women in Aviation is designed to unify a small population dispersed across a wide geographic area and a variety of professional pursuits; it aims to raise consciousness, to serve informational needs, to facilitate self-help and networking, and to encourage greater accomplishment. Judging by circulation figures and similar indicators, this unique initiative has been a success. In arranging financing for this venture, Carmien had to struggle with estimates of potential clientele. This was very difficult to do, in part because individual women are broadly scattered throughout the aviation community. Many people felt that the total population of potential readers, and hence advertisers, was too small. However, four years after its founding, *Women in Aviation* is still going strong and is paying its bills. Its subscription list remains modest, around two thousand, but it is continuously growing. The number of pages has increased, but manuscripts are now submitted for publication at a rate that far exceeds the space available. While a sophisticated survey of readership is beyond current means, it is clear that the magazine is reaching its intended audience. A small number of older, mostly retired female professionals read it, as do airline pilots and air traffic controllers.

Most issues of *Women in Aviation* feature, often prominently, an article on the airwomen. Carmien is quite direct about how this fits with the magazine's major purposes: she seeks to exploit the inspirational potential of the airwomen's legacy. Just as the magazine is intended to reduce women's isolation by increasing communication across geographic distance and occupational lines, so can it do so by transcending time. If today's female aviation professionals understand that others loved the field, struggled, and succeeded before them, they can perhaps more easily meet challenges. Models for emulation and sources of inspiration, by their nature, mean that one is not alone. With such motivation on the part of the publisher, it is not surprising that in nine issues, admittedly selected by an ad hoc process, one finds articles or photographs on the following popular airwomen: Pancho Barnes, Bessie Coleman, Amelia Earhart, Ruth Law, Nancy Harkness Love, Mathilde Moisant, Harriet Quimby, Blanche Scott, Katherine Stinson, Louise Thaden, and Edna Gardner Whyte. In addition, the following lesser-known airwomen are also referenced: Janet Waterford Bragg, Helen Hodge Harris, Mary von Mach, Nadine Canfield Nagle, Nancy Hopkins Tier, and Jessie Woods. Only one of the nine issues made no reference at all to the airwomen.

Carmien, sensitive to women's history in general, believes that the emergence of appropriate leaders is crucial in women's quest for equity. She believes that in aviation, there has been too much reinventing of the wheel. That is, if today's leaders can profit from lessons of the past, their task will be much easier.

Carmien envisions a purpose beyond serving aviation professionals: socialization of young people. *Women in Aviation* is sent to high schools and colleges where she hopes it will demonstrate to young women some of the broad career options that they possess. Here again, content featuring the airwomen can play a key role. Most youngsters know of Amelia Earhart, but if she is thought of as a solitary superstar, her inspirational value is limited. A history in which many women have achieved success is both more accurate and more likely to stimulate emulation as opposed to sedentary awe. Indeed, airwomen are a key part of the most visionary women's professional aviation publication available today (personal communication with Amy Carmien, Jan.–Nov. 1992).

Although Carmien's single-handed effort shows unusual initiative, she is not alone. Others have had similar ideas and, where there has been institutional financial support, the results have verged on the spectacular. Peggy Baty, now chief academic officer at Parks College of St. Louis University, has organized an annual National Women in Aviation Conference. Baty is a successful pilot with a number of ratings. In 1989, she was serving as director of the aviation management program at Embry-Riddle Aeronautical University. Extremely successful in her own right, Baty initially had little interest in gender-specific issues in aviation. Baty's sources of inspiration in achieving her great professional success had nothing to do with the airwomen; for the most part, she, like the great majority of Americans of her generation, had never heard of them.

Baty, however, became progressively more aware of the small number of women enrolled in Embry-Riddle's aviation programs and their lack of role models. She began to think in terms of programs to inspire these people. A particular incident of gender stereotyping, in which her role at Embry-Riddle was assumed to involve flight attendant training, precipitated the decision to organize a conference. The major beneficiaries were to have been young people in general and Embry-Riddle students in particular. Calls for papers and other forms of contribution went out. The first conference, held in 1990, brought home three points. First, certain potential participants wanted to call attention to the airwomen, who were regarded as fine examples of the values that the conference was designed to promote. Ultimately, arrangements were made to feature prominent WASPs at the conference and Bobbi Trout was engaged as keynote speaker. Second, a great many people were interested in the conference; initial arrangements for one hundred attendees were severely strained when more than 175 showed up. Third, while the most prominent form of communication to occur at the conference was formal presentations to a largely student audience, an unanticipated form of information exchange developed as well. Since there had never been such an assemblage of women with diverse interests, experiences, and views, the conference presented a unique opportunity for networking between and among presenters and audience alike.

This brought forth immediate plans for a second conference with slightly altered goals: female participants in aviation were to be the primary targets and service to professional and other needs was the major purpose. Socialization of young women was retained as an important goal, and provision of a message for the general public was recognized as a third. This focus did not reduce the demand for programs on the airwomen. Indeed, at the 1991 conference special recognition was given to the WASPs. In addition, six of the twenty-one specialized concurrent sessions dealt in one way or another with the airwomen. Mixed in with panels on planning for job interviews, career choice dynamics, and pregnancy and the workplace were panels on, for example, Phoebe Omlie and, inevitably, Amelia Earhart.

The 1992 conference was dramatically successful, with more than five hundred participants and some sixteen commercial exhibitors. The growing importance of the conference was shown by its successful recruitment of well-known test pilot Chuck Yeager as keynote speaker. Several aviation publications either gave the conference advance billing or published reports on it. The mass media were also intrigued, especially when they learned that the conference provided a forum for female military pilots and those who, in the wake of military action in the Persian Gulf, advocated combat roles. Despite the expanded emphasis on networking (many sessions were without agenda, set aside for this purpose), the airwomen remained prominent topics for discussion. At least four panels explicitly addressed female pilots of their era.

The fourth conference will be similar in scope to the third, though Baty plans to broaden the educational function. She intends to promote the conference among school teachers, urging attendance at special sessions on the weekend. The explicit focus on career choices for young women will be supplemented by panels on aviation history. The latter, like the schoolteacher workshops Baty now conducts apart from her conference responsibilities, will reference the airwomen.

Baty seems well aware that she has tapped into something with great potential. Inevitably, she had to face the question of whether she was competing with the Ninety-Nines, the organization of female pilots that dates back to the twenties. Like other observers of this group's role over the years, Baty notes that the Ninety-Nines are in large part a club dedicated to pleasant fellowship and social events. The conference, by contrast, clearly serves professional advancement needs. Some participants have called for the formation of an aggressive, permanent organization to parallel the conference. The need for sustained services, for continuous and systematic information exchange, for publicity and positive public relations, and even for political action and lobbying is strongly felt by many women in aviation. For a variety of good reasons, Baty is reluctant to establish a formal association, but she recognizes that the issue is on the table (personal communication with Peggy Baty, March–Dec. 1992).

The paragraphs above recall some of the points made in Chapter 4: by the late thirties and early forties, the "demonstration" stage of women's involvement in aviation had run its course; having shown that they possessed the necessary abilities, it was time for women to establish their place in the aviation professions. Unfortunately, the leadership for this new stage did not emerge and an appropriate organizational infrastructure did not develop. Now, more than a half century later, it would appear that the demand for professional support is reaching levels that cannot be ignored. Will women in aviation at last enthusiastically embrace the professional stage?

Whatever its agenda turns out to be, it appears that the Women in Aviation Conference needs the memory of the airwomen. This memory — of accomplishment, daring, and courage — is an important part of mobilizing female pilots for professional action and educating young people. Beyond this, perhaps conference organizers will consider the sad lesson of post–World War II airwomen unserved by organizations that never developed.

Educating Young People

While both Carmien and Baty have service to the women's aviation community as a primary goal, they have acquired an educational mission as well. The Women in Aviation Conference in particular has deep educational roots. More explicit efforts at educating the younger generation are under way as well. Many of these are of very recent origin, but two deserve particular mention. During the mid-eighties, a curriculum specialist from Minnesota, Cheryl Young, almost by accident discovered the airwomen. In the process of working on the problem of integrating women's history into the middle school curriculum, she happened on some materials, including Keil's book on the WASPs, dealing with the air-women. Though Young had no previous knowledge of these pilots, nor connection with aviation for that matter, she was intrigued and quickly incorporated segments on the WASPs into her own middle school teaching efforts. By 1988, Young felt she needed to make contact with living WASPs if she was to achieve the highest degree of realism in her teaching. A few calls led her to Elizabeth Strohfus, a lifelong Minnesota resident. Strohfus had served as an instrument instructor and a gunnery target tow-plane pilot in the WASPs. Strohfus sought a job with Northwest Airlines after the war, but, like so many others, she was turned down.

Forty-seven years passed before anyone asked her about her military career — much less brought up the importance of leaving a legacy.

Through mutual reinforcement, Young and Strohfus became strong public advocates of the WASPs' story. Young in general believed that women should have many options as to how to contribute to society; she felt that knowledge of the WASPs would greatly broaden the horizons of young people. Strohfus came to believe that her story could inspire greater women's achievements in the future.

The two have devised plans for two children's books, a general recounting of the WASP experience, and Strohfus's personal story. These materials are still in preparation, as are a series of multimedia curriculum units featuring these airwomen. Such units will be available for young people from the elementary grades through high school. Young envisions them as circulating to a great many schools from a few central educational materials centers.

The most dramatic outcome of this partnership has been a long series of personal appearances and slide shows. Strohfus made a few presentations to local clubs and history buffs. These proved immensely popular and demand mushroomed. It became necessary to move from simple slides to professionally produced presentations. Strohfus developed a very effective, engaging style. Word of the WASP presentation spread beyond the local areas to diverse new clientele. Audiences grew larger as the pair appeared on college campuses and before groups of business executives — including, ironically, those of Northwest Airlines. Especially gratifying was the demand from schools; Strohfus and Young have made presentations to single classes of second graders and to complete high schools. Strohfus has spoken before tens of thousands of children, of all ages. The media, including large metropolitan dailies, have taken note. So, incidentally, have local B. Dalton bookstores; they have engaged her for in-store appearances to promote aviation book sales. Recently, national aviation groups have begun to engage Strohfus on a regular basis; most notably, she was a featured speaker at the 1992 annual fly-in of the massive Experimental Aircraft Association.

Strohfus is not unique in finding herself in the public eye. The WASPs, as virtually the last living airwomen, are especially likely to be objects of this newfound attention. Not only have individual WASPs written books, but they have shown a renewed pride in many other ways as well. Most notably, their organization has been revitalized. Almost moribund until the late seventies, the group now elects officers and sponsors national reunions. Many WASPs, such as 1992 president Yvonne (Pat) Pateman, find themselves in great demand as speakers for organizations and schools.

Pateman is very pleased that the WASPs are frequently called upon by the Women Military Aviators, an organization composed of female pilots who currently fly in the Air Force, Navy, Army, or Coast Guard. Without hesitation they acknowledge the WASPs as role models and incorporate them into their organization. Even the military now pays solicitous attention to the WASPs; all the attendees at the 1992 reunion, held in San Antonio, were invited as honored guests to Randolph Air Force Base where they enjoyed a special flight in a Lockheed C-5A transport plane. The base commander at the Corpus Christi Naval Air Facility has arranged, on a permanent basis, for a WASP to be a part of the ceremony at which new naval aviators, men and women alike, receive their wings.

Attention and pride go hand in hand. The WASPs have finally come to the point where they collectively believe in themselves and for the first time they are aggressively seeking to leave a legacy. One of the most interesting parts of this effort is the endowment of scholarships for young women who seek aviation or aerospace careers. The exact format of this college-level award has not yet been determined, but many WASPs are either contributing significant amounts directly to the endowment or are heavily involved in raising funds from outside sources (personal communication with Yvonne Pateman, Sept. 1992). Nothing of this sort has happened before in the history of the WASPs.

Educating the Public at Large

While both Strohfus and the WASPs in general may have youth as their primary target, they have increased the knowledge and changed the attitudes of other segments of the population as well. At least one organization has deliberately taken a very broad educational aim; in 1979, about the time sustained written works about the airwomen started once again to appear, two members of the Ninety-Nines began to formulate plans for the International Women's Air and Space Museum (IWASM). Beatrice Steadman and Kate Schamberger were increasingly concerned that women's role in the history of aviation was largely unknown. Moreover, such materials as were generally available were often in error. The two hoped that the Ninety-Nines might be a driving force in correcting this situation, perhaps through a new approach to its archives program.

When this proved impossible, a separate, tax-exempt corporation was chartered in Ohio. Money was raised, people were organized, materials were collected, and physical facilities were arranged. The doors of a new museum were opened in 1986. Over time, the mission of accumulating accurate records for posterity was augmented by an aggressive program of positive information dissemination. The corporation developed into a complex institution with community programs and provisions to make its collections available to researchers in the field. It became more than a repository for records of the past; it came to view itself as a recorder of women's history in the making as well. Sophisticated educational programs developed, including the production of topical videos, the organization of school field trips, the sponsorship of lectures by prominent women in the aviation community, and the development of programs for high school career fairs.

The focus of the museum is very broad. The space program, for example, is prominently featured in most of what it does. However, so are the airwomen. It was, after all, their relative obscurity that prompted the founding of the museum in the first place. A major portion of the exhibit space features the airwomen, particularly — given artifact availability — the later ones. Indeed, the president of the corporation is Nancy Hopkins Tier, an established if less well known airwoman of the late thirties. At the age of eighty-three, she still carries on an active outreach program on behalf of the museum. Her presentations are much in demand with organizations nationwide. WASP Nadine Canfield Nagle devotes considerable time to the museum and has similarly offered many public service programs to community groups.

Although at first IWASM did not receive much in the way of cooperation from other museums — no one was interested in its offers of travelling exhibits, for example — it proved to be an idea whose time had come. The number of visitors has increased steadily. IWASM's success has attracted the attention of many women in the aviation community. For example, the Association of Women Airline Pilots now maintains a permanent exhibit at the museum, featuring information placards, photos, and items such as insignia and uniforms.

Most impressive perhaps is the fact that the museum will soon leave its present modest quarters, a house in Centerville, Ohio, once owned by the Wright family. The city of Dayton has ceded four and one-half acres of prime downtown property for the construction of a new, expanded museum. A capital campaign to support construction of the thirty-thousand–square-foot facility is in progress as of 1993. The new structure will be located near the original Wright brothers bicycle shop, which is undergoing restoration as a national monument. Indeed, there will be a major aviation historic park in Dayton. Because of the Wrights and the many aviation events that have happened there, it is hardly surprising that Dayton will be the site of such an operation; that the International Women's Air and Space Museum will be a prominent part of it is a sign that things have changed. In the thirties a women's museum might have been part of a major public monument to aviation — it would have been fully consistent with the attention and positive commentary that the airwomen enjoyed in the press. In the fifties, this would have been impossible. In the seventies, a few would have found the idea novel and intriguing. That it is a reality in the nineties bodes well for the legacy of the airwomen.

Signs in Popular Culture

Most of the activity discussed thus far in this chapter is quite purposeful; that is, it was undertaken by people who want to make a statement, who have a message of which the airwomen are an important part. That so many people have such statements to make is of great significance. Even more telling, perhaps, is the increasing secondary activity that in some way involves the airwomen. If the culture is truly receptive to the memory of the airwomen, entrepreneurs will seek to capitalize on this fact quite in the absence of any particular pro-airwomen motivation. For example, it is quite possible that, in contrast to museum officers, few city officials in Dayton feel a strong inherent need to communicate the airwomen's story. Their interest in having an expanded IWASM in the downtown area may be more easily explained in terms of making the city more attractive to tourists, more livable, and, as a consequence, more economically and socially viable.

There are other indicators that the airwomen have influenced this kind of thinking. One of the most interesting examples occurred in Sweetwater, Texas, the location of Avenger Field, the WASP training facility and the world's only all-women airbase. During World War II, Sweetwater welcomed the WASPs warmly, just as many other communities supported military personnel stationed nearby. Indeed, the town's reputation for hospitality had begun earlier, when British and American airmen had been stationed there. With disbandment, however, the WASPs faded from community consciousness and the airfield was put to civilian use.

For a quarter of a century, the memory of the WASPs lay dormant. In 1968, plans were revealed for closing the radar base that had occupied a portion of the airfield after World War II. This caused some consternation in Sweetwater, as the base had been an important contribution to the economy and the airfield property threatened once again to lapse into disuse. The chamber of commerce and other community leaders intensified their promotion of Sweetwater municipal airport, formerly Avenger Field, as the site for a new state community college. Attention to the airport property seemed to revive thoughts of the WASPs. A search for artifacts revealed that all tangible signs of the WASPs had disappeared, except the shell of a stone wishing well. This well had been of great ceremonial importance during the WASP years; successful program completion was always followed by a ritual dunking.

Largely through the leadership of Rigdon Edwards, a WASP flight instructor during the war, the community began a long-term effort to reassert its connection to the airwomen. The chamber's first effort was to organize a national reunion of the WASPs. This was particularly significant because the WASPs had been essentially dormant, as a unit, since 1944, lacking both organization and pride. It is not surprising that the impetus for revitalization of the WASPs had to come from an external source. The effort was well orchestrated, however, and the WASPs responded. Ultimately, the first of a long string of reunions was held in Sweetwater in 1972.

The city also renamed the municipal airport Avenger Field, returning to the designation of the WASP era. Arrangements for the permanent protection of the wishing well were put in place, but progress on the ultimate recognition of the WASPs, a substantial permanent monument, was slow. This project has really taken off only in the last few years. The key features of the monument are to be a life-size bronze statue of a WASP in full flying attire and a series of black granite plaques bearing the names of all the women who trained at Avenger Field. Some $200,000 has been raised, most of it locally. The sculptor, Dorothy Swain Lewis, has likewise donated her services. As the land for the site was already in city hands, a very respectable start has been made and the monument is scheduled for dedication in May 1993. This is quite a remarkable effort for a town of only twelve thousand inhabitants.

Although many in the community are motivated by nostalgia or the sincere desire to recall a heroic past, the civic leaders of Sweetwater are not unaware of the monument's potential role as a tourist attraction. Indeed, they are pursuing the designation of Avenger Field as a National Historic Site with the thought of increasing the overall draw. The monument is a singular phenomenon. For decades, the WASPs were largely forgotten in this community, and nearly all physical evidence of their presence had disappeared. What is happening in Sweetwater today clearly shows that as far as the legacy of the airwomen is concerned, we have entered a new era.

There are other popular signs of the reemergence of the airwomen. In 1983, Blanche Scott appeared on a U.S. postage stamp, appropriately enough an airmail issue. In 1991, a similar number bore the likeness of Harriet Quimby. Jacqueline Cochran is scheduled to join this pair in the near future. To be sure, given the large number of denominations and the tradition of frequent and various issues, it is hardly a sign of public prominence to be shown on a stamp. The point is, however, that no airwomen, save the ubiquitous Amelia Earhart in 1968, has ever before been recognized in this fashion. Whatever forces impel the movers and shakers in postage stamp design processes, they are different from before. Airwomen, for the first time since the thirties, may be important in the promotion of commercial sales. One intrepid entrepreneur has recently begun marketing tote bags and T-shirts with a charming picture of an unidentified airwoman in an open cockpit. Reportedly, demand swamped her 1992 sales operation at the Experimental Aircraft Association's Oshkosh airshow.

Conclusion

In a word, the airwomen are back. Why a resurgence of interest now, rather than ten, twenty, or thirty years ago? Undoubtedly part of the answer lies in a general contemporary appetite for nostalgia — a phenomenon not unusual in times of economic adversity and social uncertainty. Also, it is quite possible that the airwomen are riding the coattails of the increasing number of contemporary professional female pilots. As Chapter 4 demonstrated, the airwomen were not sources of inspiration for these new female pilots. Nonetheless, some of them are quite familiar with the airwomen, presumably acquiring that knowledge in a search for roots. Relatedly, the travelling public cannot avoid being aware of professional female pilots. This is a new phenomenon and as such arouses a certain amount of public curiosity, including an interest in the background of it all.

A comprehensive explanation of the reemergence of the airwomen is beyond the scope of this book; rather, this book aims to underscore the opportunity that it presents. The major lesson of the airwomen has been in the form of a warning: an opportunity for women's advance in this culture was missed. It appears, however, that the story does not end there. What was so conspicuously lacking in the decade following World War II — a popular group of early aviators to admire and perhaps emulate — is, in fact, emerging. This particular episode of obscurity, silence, and neglect is coming to an end.

Many individuals want to spread word of the airwomen's legacy. Well-run organizations are in place, permanent monuments are under construction, and communication networks are well established. Overall, there has been a rather substantial capital investment. Publishers, educators, public officials, and entrepreneurs have undertaken to preserve this legacy in a variety of ways. The airwomen stand on a new threshold. That they were heroes there is no doubt. That we have learned from their story is certain. That their emerging story can benefit our culture is beyond question. There is hope for a legacy after all.

Glossary

A and P rating — An official Federal Aviation Administration designation certifying an individual as an airframe and powerplant mechanic. The holder of such a rating must demonstrate certain competencies and pass appropriate tests. Only persons with this rating may perform repair and maintenance operations on normally certified U.S. aircraft.

aerobatics — A contraction of "aerial acrobatics." The sport of performing airplane maneuvers that are unusual or beautiful in execution. Such maneuvers take great skill and precision. Aerobatics may be performed for the entertainment of airshow audiences or in formal, judged competitions.

air marking — The practice of painting large navigational symbols on rooftops to serve as visual aids for pilots. Undertaken in the 1930s, generally obsolete today.

amphibian — A type of airplane that is capable of operating from both land and water. Equipped with wheels, as well as either a hull or floats.

ATA girl — A female pilot in the British Air Transport Auxiliary. This unit was formed near the beginning of World War II. Its purpose was to organize those pilots who were not considered candidates for combat roles for maximum effectiveness in the defense effort. Most of the piloting tasks, as the name of the unit implies, involved transporting people, materiel or aircraft. Both men and women were employed in the ATA. The reference to the female pilots as "girls," sexist though it is in contemporary usage, was accepted at the time; this book retains the use of this standard term.

autogiro — A type of aircraft that derives much of its lift from a large overhead rotor, rather than fixed wings. An early precursor to the helicopter; now obsolete for most purposes.

AWTAR — All-Woman Transcontinental Air Race. A handicap event for female pilots with all types of planes. An annual event from 1949 to 1977.

B-17, B-18, etc. — A designation code used in World War II for particular types of Army aircraft. The "B" stood for "bomber." Such airplanes were designed to drop explosives on enemy targets. They were typically large and complex and generally had at least two engines. The number in the code refers to the sequence in which the Army obtained the various types. The B-1 was a huge, ungainly Barling biplane built in 1925, while B-52s are still in service today.

barnstorming — An early form of aviation business enterprise that usually involved a single individual and a single airplane. The barnstormer, capitalizing on the novelty and romance of flight, flew from town to town to give exhibitions and offer airplane rides. The term probably derives from the fact that landing fields in these preairport days were often pastures or fields in the proximity of barns.

biplane — An airplane with two sets of wings, one mounted above the other. This configuration was the standard in the early days of aviation. The large total wing area produced high amounts of lift, an advantage with low-powered engines and short landing fields. A major disadvantage was that this configuration generated considerable drag, which resulted in lower speeds. Biplanes, now obsolete for most functions, are also structurally more complex than monoplanes, because the latter have only a single set of wings.

C-54, C-55, etc. — A designation code used in World War II for particular types of Army aircraft. The "C" stood for "cargo." Such airplanes were designed to carry materiel or military personnel from place to place, and thus typically were rather large. Often they were militarized versions of planes originally designed as commercial airliners. The number in the code refers to the sequence in which the Army obtained the various types. A similar "C" code is used to designate Air Force heavy haulers today.

CAA — See Civil Aeronautics Authority.

cantilever wing — An aircraft wing that is supported only by fittings where it attaches to the fuselage. Brace wires or struts are unnecessary, resulting in lower drag and higher speeds. Almost all modern aircraft wings are of cantilever design, but most early designs, especially biplanes, were not.

Civil Aeronautics Authority — A U.S. governmental agency that evaluated, regulated, and promoted civilian aviation. The precursor to today's Federal Aviation Administration (FAA).

closed-course racing — Competition between airplanes, all of which follow a designated course laid out as a loop. This course is analogous to a race track for horses or cars and a given race may require several laps. The loop may assume any of several shapes, but it is generally a polygon rather than an oval or a circle. The corners of the polygon are marked with prominent pylons; hence this activity is also called pylon racing. Speed records are sometimes attempted over closed courses, and in this case only one plane participates, racing against the clock.

dirigible — An airship characterized by a large, streamlined container for lighter-than-air gas and some means of propulsion. Dirigibles were popular in the twenties and thirties for both passenger transport and naval patrols. They were similar to today's blimps, except that they were larger and had rigid structures to enclose the gas bags.

ditching — The act of making an emergency landing in a body of water with an aircraft equipped only for land operations. A successful ditching involves minimal damage and escape from the airplane before it sinks.

divebombing — A technique in which an explosive is delivered by diving an airplane directly at a target. One aims the airplane such that when the bomb is released, it follows the established trajectory to the target while the airplane alters course and flies away. Level bombing, by contrast, involves release of the explosive from an airplane at high altitude in level flight, usually with the aid of sophisticated optical sighting devices.

endurance flying — The practice of continuously staying aloft in an airplane for the longest possible time. Usually done for the purpose of testing components or setting records.

FAI — See Federation Aeronautique Internationale.

FBO — See fixed-base operator.

Federation Aeronautique Internationale — An international organization that keeps records for various aircraft categories of aircraft performances and competitions. Headquartered in Paris, it is recognized worldwide as the repository of official documentation on these matters.

fixed-base operator — Businesses that provide fuel, maintenance, and other aviation-related services at airports. FBOs often provide flight instruction and hangar rental as well. The origin of the term is obscure, but it probably derives from the fact that entrepreneurs providing such services were necessarily confined to given airports, whereas other early aviation businesspeople, such as barnstormers, were likely to be wanderers. Indeed, barnstormers were originally called gypsies.

flying boat — A type of seaplane, an airplane designed to take off from and land on water. The lower portion of a flying boat's fuselage is shaped like a boat hull; it actually contacts the water and provides the flotation. Other seaplanes use separate pontoons or floats, which keep all basic components well above the water.

general aviation — Civilian aeronautical activities undertaken for business or recreation. This category specifically excludes commercial airlines and the military. In general, the "light planes" of the country are in the general aviation fleet.

ground-loop — A violent lateral deviation from an airplane's intended path as it moves on the ground. Ground-loops usually occur during takeoff or landing, and they result from loss of control. A mild ground-loop involves nothing more than running off the runway at an angle. In a severe incident, the airplane's path may describe a complete circle, hence the term ground-loop. Wings, propeller, or other portions of the airplane may touch the ground, causing damage. Collapse of the landing gear is also likely.

LORAN — An electronic navigation system that allows the determination of an aircraft's geographic location with a high degree of accuracy. It is based on the simultaneous reading of the compass bearing of widely separated stations, each of which broadcasts a unique signal.

P-35, P-36, etc. — Designation code used in World War II for particular types of Army aircraft. The "P" stood for "pursuit," as the class was designed to pursue intruding enemy aircraft and shoot them down. Aircraft with this designation were fast, maneuverable, and relatively small, typically with only one engine. They were often called "fighters," and in fact the "P" designation was eventually changed to "F." The number in the code refers to the sequence in which the Army obtained the various types. The first in the series was an open cockpit biplane, the Curtiss P-1 of 1926; the last airplane produced in any quantity with the "P" designation was the P-80, an early jet that first saw service in 1945.

pylon racing — See closed-course racing.

supercharging — A practice that improves the performance of engines by compressing the air-fuel mixture prior to combustion. Similar to current automobile turbo-charging. Especially useful on airplanes, because engine performance is otherwise degraded as one climbs to altitudes where atmospheric density is significantly lower than at sea level. Piston engines that drive propellers may be supercharged; the term does not describe jet or other turbine technology.

T-6, T-7, etc. — Designation code used in World War II for particular types of Army aircraft. The "T" stood for "trainer." Such airplanes were designed as instructional vehicles for pilots or other aircraft crew members. They varied greatly, depending on the particular tasks for which students were being trained and the degree of mastery that the students possessed. The number in the code refers to the sequence in which the Army obtained the various types. Many military aircraft now in service still use the "T" designation.

VOR — An electronic navigation system featuring a series of stations, each of which broadcasts a signal on a given frequency. Aircraft equipped with proper receivers can determine the compass bearing to any station by tuning to the proper frequency. It is possible to track to any of these stations and fly directly over them simply by reference to an instrument face.

WAFS — Women's Auxiliary Ferrying Squadron. The first U.S. paramilitary organization of female pilots. Formed through the efforts of Nancy Harkness Love in 1942. Later absorbed by the WASP program.

warbird — An informal reference to an aircraft designed for the armed services. Most often used to describe older aircraft no longer in actual military service, for example those in the hands of collectors or those converted to civilian uses such as racing.

wing loading — The amount of weight that must be borne, and hence the amount of lift that must be generated, per unit of area of an airplane's wing. Low wing loading aircraft have relatively large wings for their weight, while those with high wing loadings have relatively small surfaces. Everything else being equal, high wing loading aircraft must travel faster in order to generate the necessary lift and this, in turn, requires higher takeoff and landing speeds; such aircraft are often regarded as more difficult to fly.

wing walking — A type of airshow act where a person leaves the cabin or cockpit while an airplane is in flight. In an apparently precarious position, this person literally walks on the wings or does other seemingly dangerous things. Usually performed at low altitude to afford spectators a clear view. Prominent in the twenties and thirties, wing walking survives as a form of entertainment today. Wing walkers favor biplanes because struts and braces afford convenient handholds.

References

Ackerman, Diane. *On Extended Wings* (New York: Atheneum, 1985).

Adams, Jean, and Margaret Kimball. *Heroines of the Sky* (Garden City, NY: Doubleday-Doran, 1942).

"Alaska Crash." *Airwoman* 2 (Sept., 1935): 15.

Allen, Oliver E. *Airline Builders* (Alexandria, VA: Time-Life Books, 1981).

"Amelia Earhart." *New York Times* (July 20, 1937): 24.

"American Supergirl and Her Critics." *Literary Digest* 95 (Oct. 29, 1927): 52–57.

Anderson, Karen. *Wartime Women: Sex Roles, Family Relations, and the Status of Women During World War II* (Westport, CT: Greenwood Press, 1981).

"Auriol Breaks Sound Barrier at 687.5 m.p.h." *New York Times* (Aug. 30, 1953): 1.

"A Woman Flies 590 Miles." *New York Times* (Nov. 20, 1916): 12.

Backus, Jean L. *Letters from Amelia* (Boston: Beacon Press, 1982).

Batten, Jean. *Alone in the Sky* (Shrewsbury: Airlife Publishing Co., 1979).

Batten, Jean. *My Life* (London: George C. Harap, 1938).

Baty, Peggy. "Those Wonderful Women in Their Flying Machines," *On Approach* (Spring 1990): 8–9.

Baxter, Gordon. "Iron Edna," *Flying* 111 (May 1984): 108.

Beatty, John. "I Let My Daughter Fly," *American Magazine* 130 (Sept. 1940): 32–33.

Bell, Elizabeth S. "The Women Flyers: From Aviatrix to Astronaut," in Pat Browne (ed.) *Heroines of Popular Culture* (Bowling Green, OH: Bowling Green State University Popular Press, 1987), 54–62.

Berkin, Carol Ruth, and Mary Beth Norton (eds.). *Women of America: A History* (Boston: Houghton-Mifflin, 1979).

Bernheim, Molly. *A Sky of My Own* (New York: Rinehart, 1959).

Bilstein, Roger E. "The Airplane, the Wrights, and the American Public," in Richard P. Hallion (ed.), *The Wright Brothers* (Washington: National Air and Space Museum, 1978), 3–51.

Bilstein, Roger E. *Flight in America 1900–1983* (Baltimore: Johns Hopkins University Press, 1984).

References

Bilstein, Roger E. *Flight Patterns* (Athens: University of Georgia Press, 1983).

Boase, Wendy. *The Sky's the Limit* (New York: Macmillan, 1979).

Boom, Kathleen Williams. "Women in the AAF," in Wesley F. Craven and James L. Cate (eds.), *The Army Air Forces in World War II, Vol. 7* (Chicago: University of Chicago Press, 1958), 525–550.

Brennen, T. C. Buddy. *Witness to the Execution: Odyssey of Amelia Earhart* (Frederick, CO: Renaissance House, 1988).

Briand, Paul. *Daughter of the Sky* (New York: Duell, Sloan, and Pierce, 1966).

Brooks-Pazmany, Kathleen L. *United States Women in Aviation — 1919–1929* (Washington, D.C.: Smithsonian Institution Press, 1983).

Brown, Dorothy M. *Setting a Course: American Women in the Twenties* (Boston: Twayne Publishers, 1987).

Brown, Marjorie. "Flying Is Changing Women," *Pictorial Review* 31 (June 1930a): 30.

Brown, Marjorie. "What Men Flyers Think of Women Pilots," *Popular Aviation and Aeronautics* 4 (March 1929): 62–64.

Brown, Marjorie. "Why Women Pilots Have Organized," *Aeronautics* 6 (March 1930b): 35–36.

Bruce, Mrs. Victor. *The Bluebird's Flight* (London: Chapman and Hall, 1931).

Buffington, H. Glenn. "Flying Life of Viola Gentry," *American Aviation Historical Society Journal* 13 (Summer 1968a): 125–126.

Buffington, H. Glenn. "Jean LaRene: Professional Pilot of the Golden Age," *American Aviation Historical Society Journal* 14 (Spring 1969): 47–49.

Buffington, H. Glenn. "Louise Thaden," *American Aviation Historical Society Journal* 12 (Winter 1967): 285–287.

Buffington, H. Glenn. "Phoebe Fairgrave Omlie," *American Aviation Historical Society Journal* 13 (Fall 1968b): 186–188.

Buffington, H. Glenn. "Ruth Nichols and the Vegas," *American Aviation Historical Society Journal* 11 (Spring 1966): 34–37.

Burnham, Margaret. *The Girl Aviators and the Phantom Airship* (New York: Hurst and Co., 1913).

Burnham, Margaret. *The Girl Aviators on Golden Wings* (New York: Hurst and Co., 1911).

Burnham, Margaret. *The Girl Aviators' Motor Butterfly* (New York: Hurst and Co., 1912).

Burnham, Margaret. *The Girl Aviators' Sky Cruise* (New York: Hurst and Co., 1914).

Callen, Charles L. "There's No Stopping a Woman With Courage Like This," *American Magazine* 108 (August 1929): 28–29.

Chafe, William Henry. *The American Woman: Her Changing Social, Economic, and Political Roles* (New York: Oxford University Press, 1972).

Chafe, William Henry. *The Paradox of Change* (New York: Oxford University Press, 1991).

References

Chase, Gene. "Louise Thaden in Retrospect," *Sport Aviation* 39 (Sept. 1990): 12–19.

Chun, Victor K. "The Origin of the WASPs," *American Aviation Historical Society Journal* 14 (Winter 1969): 259–262.

Cleveland, Reginald M. "Air Travelers Ride in Comfort," *New York Times* (Feb. 24, 1935): VII-15.

"Clouds and Victory." *New York Times* (July 24, 1937): 14.

"Cochran Becomes First Woman to Break Sound Barrier." *New York Times* (May 19, 1953): 32.

Cochran, Jacqueline. *The Stars at Noon* (Boston: Atlantic-Little Brown, 1954).

Cochran, Jacqueline, and Mary Ann Bucknum Brinley. *Jackie Cochran* (New York: Bantam Books, 1987).

"Cochran Sets 2 More Speed Marks." *New York Times* (June 4, 1953): 8.

Cole, Jean Hascall. *Women Pilots of World War II* (Salt Lake City: University of Utah Press, 1992).

Corn, Joseph J. "Making Flying Thinkable: Women Pilots and the Selling of Aviation, 1927–1940," *American Quarterly* 31 (Fall 1979): 556–571.

Corn, Joseph J. *The Winged Gospel* (New York: Oxford University Press, 1983).

Courtney, W. B. "High Flying Ladies," *Collier's* 90 (Aug. 20, 1932): 29ff.

Cranston, Claudia. *Sky Gypsies* (Philadelphia: J. B. Lippincott, 1935).

Crouch, Tom D. *A Dream of Wings: Americans and the Airplane, 1875–1905* (Washington: Smithsonian Institution Press, 1989).

Crowley, Kitty. *First Women of the Skies* (New York: C.P.I., 1978).

Dailey, Janet. *Silver Wings Santiago Blue* (New York: Pocket Books, 1984).

"Death Ruled Suicide." *New York Times* (Oct. 20, 1960): 39.

deLeeuw, Hendrik. *From Flying Horse to Man in the Moon* (New York: St. Martin's Press, 1963).

"Designs Flying Clothes." *New York Times* (Nov. 24, 1933): 30.

Devine, Thomas E., with Richard M. Daley. *Eyewitness: The Amelia Earhart Incident* (Frederick, CO: Renaissance House, 1987).

"Domestic Aviation." *New York Times* (Nov. 26, 1944): IV-8.

Donahue, James A. *The Earhart Disappearance* (Terre Haute, IN: Sunshine House, 1987).

Douglas, Deborah G. *United States Women in Aviation 1940–1985* (Washington, D.C.: Smithsonian Institution Press, 1991).

Douglas, Deborah G. "WASPs of War," *Aviation Heritage* 2 (May 1992): 46–53.

"Do Women Hoodoo Transatlantic Flights?" *Literary Digest* 108 (Feb. 14, 1931): 32.

Drake, Francis Vivian. "Air Stewardess," *Atlantic* 151 (February 1933): 185–193.

Dwiggins, Don. *They Flew the Bendix Race* (Philadelphia: J. P. Lippincott, 1965).

Earhart, Amelia. "Flying Is Fun," *Cosmopolitan* 93 (Aug. 1932a): 38–39.

Earhart, Amelia. *The Fun of It* (New York: Brewer, Warren, and Putnam, 1932).

Earhart, Amelia. *Last Flight* (New York: Harcourt-Brace, 1937).

Earhart, Amelia. "Mother Reads as We Fly," *Cosmopolitan* 90 (Jan. 1931a): 17.

Earhart, Amelia. "The Man Who Tells the Flier, 'Go,' " *Cosmopolitan* 86 (May 1929a): 78–79.

Earhart, Amelia. *Twenty Hours, Forty Minutes* (New York: G. P. Putnam's Sons, 1929).

Earhart, Amelia. "Why Are Women Afraid to Fly?" *Cosmopolitan* 87 (July 1929b): 90–93.

Earhart, Amelia. "Women and Courage," *Cosmopolitan* 93 (Sept. 1932b): 147–148.

Earhart, Amelia. "Your Next Garage May House an Autogiro," *Cosmopolitan* 91 (Aug. 1931b): 58.

"Earhart Gets Advice on Babies; Explains It Is Due to Misunderstanding of Husband's Remarks." *New York Times* (March 5, 1935): 21.

Editors of Time-Life Books, *Bush Pilots* (Alexandria, VA: Time-Life Books, 1983).

Emme, Eugene M. (ed.). *Two Hundred Years of Flight in America: A Bicentennial Survey* (San Diego: American Astronautical Society, 1977).

"Enters Bendix Race." *New York Times* (Aug. 25, 1946): II-11.

"Entries and Finishers." *New York Times* (Aug. 31, 1947): 3.

"Fair Play for Women Fliers." *New York Times* (Nov. 22, 1927): 28.

Farmer, James H. *Celluloid Wings* (Blue Ridge Summit, PA: TAB Books, 1984).

"Fatal Accident in Boston." *Scientific American* 107 (July 13, 1912): 27.

"Feminine Flyers." *Newsweek* 18 (Dec. 1, 1941): 45.

"Feminists Stirred Over Woman Flier." *New York Times* (Nov. 8, 1935): 25.

"Florence Lowe Barnes." *Aero Digest* 17, no. 2 (July 1930): 78.

"438 m.p.h. Mark for Propeller Plane." *New York Times* (Dec. 20, 1949): 7.

Freedman, Estelle. "The New Woman: Changing Views of Women in the 1920's," *Journal of American History* 61 (Sept. 1974): 372–393.

Fuller, Curtis. "Betty Skelton Flies an Airshow," *Flying* 44 (Feb. 1949): 36ff.

Gee, Henrietta. "Ruth Law, Pioneer Woman Flyer," *New York Sun* (July 12, 1932): 23.

Gentry, Viola. *Hangar Flying* (Chelmsford, MA: Privately published, 1975).

"Gets Rumanian Air Award." *New York Times* (Oct. 21, 1939): 17.

George, Weston. "Beauty and the Bleriot," *Aviation Quarterly* 6 (First Quarter 1980): 56–73.

Gillies, Betty Huyler. "Autobiography." Signed typescript in National Air and Space Museum library file (May 24, 1970).

"Girl Pilot Is Punished: Her Sentence Is Mild." *New York Times* (Oct. 27, 1928): 18.

Goerner, Fred G. *The Search for Amelia Earhart* (Garden City, NY: Doubleday, 1966).

References

Gordon, Arthur. *The American Heritage History of Flight* (New York: American Heritage Publishing Co., 1962).

Gould, Bruce. "Milady Takes the Air," *North American Review* 228 (Dec. 1929): 691.

"Government Test Pilot Makes New Women's Record." *New York Times* (Aug. 29, 1947): 7.

Gower, Pauline. *Women with Wings* (London: John Long, 1938).

Granger, Byrd Howell. *On Final Approach* (Scottsdale, AZ: Falconer Publishing Co., 1991).

Gwynn-Jones, Terry. "For a Brief Moment, the World Seemed Wild About Harriet," *Smithsonian* 14 (Jan. 1984): 112–126.

Hageb, Alice Rogers. "Romance and Valor Add Glamour to Routine of Flying Hostesses," *New York Times* (Dec. 5, 1937): VI-5.

Hallion, Richard P. *Designers and Test Pilots* (Alexandria, VA: Time-Life Books, 1983).

Harris, Barbara J. *Beyond Her Sphere* (Westport, CT: Greenwood Press, 1978).

Harris, Grace McAdams. *West to the Sunrise* (Ames: Iowa State University Press, 1980).

Harris, Sherwood. *The First to Fly* (New York: Simon and Shuster, 1970).

Hart, Marion Rice. *I Fly as I Please* (New York: Vanguard, 1953).

Hartmann, Susan M. *The Home Front and Beyond: American Women in the 1940s* (Boston: Twayne Publishers, 1982).

Heath, Lady Mary, and Stella Wolfe Murray. *Woman and Flying* (London: John Long, Ltd., 1929).

Hodgman, Ann, and Rudy Djabbroff. *Sky Stars* (New York: Athenium, 1981).

Holden, Henry M., with Lori Griffith. *Ladybirds* (Mt. Freedom, NJ: Black Hawk Publishing Co., 1991).

Holm, Jeanne. *Women in the Military* (Novato, CA: Presidio Press, 1982).

"Hostesses of the Air." *New York Times* 24 (Feb. 17, 1932).

Howes, Durward. *American Women 1935–1940* (Detroit: Gale Research Co., 1981).

Hummer, Patricia M. *The Decade of Elusive Promise* (Ann Arbor: UMI Research Press, 1979).

"Husband-Wife Teams in the Flying Game." *Literary Digest* 103 (Oct. 26, 1930): 39–40.

"Interesting People." *American Magazine* (April 1947, July 1948).

Jablonski, Edward. *America in the Air War* (Alexandria, VA: Time-Life Books, 1982).

Jablonski, Edward. *Ladybirds: Women in Aviation* (New York: Hawthorne, 1968).

Jackson, Donald Dale. *The Explorers* (Alexandria, VA: Time-Life Books, 1983).

Jackson, Donald Dale. *Flying the Mail* (Alexandria, VA: Time-Life Books, 1982).

References

James, Edward T., Janet Wilson James, and Paul S. Boyer (eds.). *Notable American Women 1607–1950*, 3 vols. (Cambridge, MA: Belknap Press, 1971).

Johnson, L. B. "Rags to Riches," *Saturday Review* 37 (Nov. 6, 1954): 18.

"Katherine Stinson in San Diego." *Aerial Age* 2, no. 13 (Dec. 13, 1915): 303.

Keil, Sally Van Wegenen. *Those Wonderful Women in Their Flying Machines* (New York: Rawson, Wade, 1979).

King, Alison. *Golden Wings* (London: C. Arthur Pearson, Ltd., 1956).

Kitchens, John W. "Army Plans to Memorialize Katherine Stinson," Army Aviation Branch History Office, Ft. Rucker, AL, 1989.

Klass, Joe. *Amelia Earhart Lives* (New York: McGraw-Hill, 1970).

Knapp, Sally. *New Wings for Women* (New York: Thomas Y. Crowell, 1946).

Krantz, Judith. *Till We Meet Again* (New York: Crown Publishers, 1988).

Laboda, Amy. "Women with Navy Wings," *Flying* 117, no. 1 (Jan. 1990): 38–52.

"Lady Heath Expected to Receive Divorce Decree." *New York Times* (March 2, 1930): II-8.

Lambert Bill. *Barnstorming and Girls* (Manhattan, KS: Sunflower University Press, 1980).

Lansing, Elisabeth Hubbard. *Sky Service* (New York: Thomas Y. Crowell, 1942).

Law, Ruth. "Let Women Fly!" *Air Travel* 1 (Feb. 1918): 250, 284.

Law, Ruth. "Telephone Interview," *New York Times* (Nov. 20, 1916): 1.

Lea, Luanne C. "Women, Aviation and the Media," *Women in Aviation* 3 (Jan.–Feb. 1992): 3–4.

Lemon, Dot. *One One* (Los Angeles: Olympic, 1963).

Leonhirth, Janene. "Tennessee's Experiment: Women as Military Flight Instructors in World War II," *Proceedings of the Second Annual National Conference on Women in Aviation*, St. Louis, MO (March 21–23, 1991).

Lindbergh, Anne Morrow. *Listen! The Wind* (New York: Harcourt, Brace, 1938).

Lindbergh, Anne Morrow. *North to the Orient* (New York: Harcourt, Brace, 1935).

Lomax, Judy. *Women of the Air* (New York: Ballentine, 1987).

Loomis, Vincent, with Jeffrey L. Ethell. *Amelia Earhart: The Final Story* (New York: Random House, 1985).

Lovell, Mary S. *The Sound of Wings* (New York: St. Martin's Press, 1989).

Lovell, Mary S. *Straight on Till Morning* (New York: St. Martin's Press, 1987).

Lowell, Edith. *Linda Carlton, Air Pilot* (Akron, OH: Saalfeld Publishers, 1931a).

Lowell, Edith. *Linda Carlton's Ocean Flight* (Akron, OH: Saalfeld Publishers, 1931).

Mackenzie, Catherine. "Hostesses of the Sky," *New York Times* (March 28, 1937): XII-1.

Macksey, Joan, and Kenneth Macksey. *The Book of Women's Achievements* (New York: Stein and Day, 1976).

Magid, Ken. *Women of Courage*. Videotape (Lakewood, CO: KM Productions, 1990).

References

"Mantz Roars 435 Miles an Hour." *New York Times* (Aug. 31, 1946): 1.

Markham, Beryl. *West With the Night* (San Francisco: North Point Press, 1983).

Martyn, T.J.C. "Women Fliers of the Uncharted Skies," *New York Times* (Aug. 10, 1930): VI-12.

May, Charles Paul. *Women in Aeronautics* (New York: Thomas Nelson, 1962).

Meyer, Dickey. *Needed: Women in Aviation* (New York: Robert M. McBride, 1942).

"Miss Katherine Stinson's Looping at Night." *Aerial Age* 2, no. 18 (Jan. 17, 1916): 424.

"Miss Quimby Flies English Channel." *New York Times* (April 17, 1912): 15.

"Miss Ruth Law Presented with $2,500 Purse." *Flying* 5, no. 12 (Jan. 1917): 497.

Moll, Nigel. "Cat Shot: Zero to 170 MPH in Three Seconds," *Flying* 119 (May, 1992): 64–77.

Moolman, Valerie. *Women Aloft* (Alexandria, VA: Time-Life Books, 1981).

"More Ocean Flights." *New York Times* (Sept. 7, 1936): 16.

Morris, Lloyd, and Kendall Smith. *Ceiling Unlimited: The Story of American Aviation from Kitty Hawk to Supersonics* (New York: Macmillan, 1953).

Morrissey, Muriel Earhart. *Courage Is the Price* (Wichita: McCormick-Armstrong, 1963).

Morrissey, Muriel Earhart, and Carol Osborne. *Amelia, My Courageous Sister* (Santa Clara, CA: Osborne Publisher, Inc., 1988).

Mosley, Leonard. *Lindbergh: A Biography* (Garden City, NY: Doubleday and Co., 1976).

Moyer, Bess. *On Adventure Island* (Chicago: Goldsmith Publishers, 1932).

"Mrs. A. Cobham Will Fly With Husband From England Around Africa." *New York Times* (Nov. 16, 1927): 3.

"Mrs. J. L. McDonnell to Go As Passenger in Race." *New York Times* (Sept. 20, 1927): 3.

Myers, Robert H. *Stand By to Die: The Disappearance, Rescue, and Return of Amelia Earhart* (Pacific Grove, CA: Lighthouse Writer's Guild, 1985).

Myles, Bruce. *Night Witches* (Novato, CA: Presidio Press, 1981).

Nevin, David. *The Pathfinders* (Alexandria, VA: Time-Life Books, 1980).

"New Crowned Queen of the Air." *Literary Digest* 53 (Dec. 2, 1916): 1485.

Nichols, Ruth. "Aviation for You and for Me," *Ladies' Home Journal* 46 (May, 1929): 9.

Nichols, Ruth. "Behind the Ballyhoo," *The American Magazine* 113 (March 1932): 43.

Nichols, Ruth. "The Sportsman Flies His Plane," *National Aeronautic Review* 8 (Apr. 1930): 33–36.

Nichols, Ruth. *Wings for Life* (Philadelphia: J. B. Lippincott, 1957).

Nichols, Ruth. "You Must Fly," *Pictorial Review* 34 (Aug. 1933): 22.

Niekamp, Dorothy Rinehart. *Annotated Bibliography of Women in Aviation* (Oklahoma City: The Ninety-Nines, Inc., 1980).

Niekamp, Dorothy Rinehart. *Women and Flight, 1978–1989* (Oklahoma City: The Ninety-Nines, Inc., 1990).

Noggle, Anne. *For God, Country, and the Thrill of It* (College Station, TX: Texas A&M University Press, 1990).

Oakes, Claudia M. *United States Women in Aviation 1930–1939* (Washington, D.C.: Smithsonian Institution Press, 1985).

Oakes, Claudia M. *United States Women in Aviation Through World War I* (Washington, D.C.: Smithsonian Institution Press, 1978).

O'Malley, Patricia. *Airline Girl* (New York: Dodd-Mead, 1944).

O'Malley, Patricia. *War Wings for Carol* (New York: Dodd-Mead, 1943).

O'Neill, Lois Decker (ed.). *The Women's Book of World Records and Achievements* (Garden City, N.Y.: Anchor Books, 1979).

O'Neill, Paul. *Barnstormers and Speed Kings* (Alexandria, VA: Time-Life Books, 1981).

O'Neill, William L. *Everyone Was Brave: A History of Feminism in America* (Chicago: Quadrangle Books, 1971).

Owen, Bessie. *Aerial Vagabond* (New York: Liveright, 1941).

Owen, Richard. "Ruth Elder's Revolt," *The Nation* 125 (Nov. 23, 1927): 569–570.

Parmalee, Deolece. "The Flying Stinson Family," *The Texas Star* (July 9, 1972): 36.

Pateman, Yvonne. "Fay Gillis Wells, Aviation Pioneer," *Aviation Quarterly* (First Quarter 1985): 48–65.

Peckham, Betty. *Women in Aviation* (New York: Thomas Nelson, 1945).

Pendergast, Curtis. *The First Aviators* (Alexandria, VA: Time-Life Books, 1981).

Pendo, Stephen. *Aviation in the Cinema* (Metuchen, NJ: Scarecrow Press, 1985).

Pickel, Margaret Barnard. "There's Still a Lot for Women to Learn," *New York Times Magazine* (Nov. 11, 1945): 14.

Piercy, Marge. *Gone to Soldiers* (New York: Fawcett Crest, 1987).

Pisano, Dominick A., and Cathleen S. Lewis. *Air and Space History: An Annotated Bibliography* (New York: Garland Publishing, Inc., 1988).

Planck, Charles E. *Women With Wings* (New York: Harper, 1942).

Poole, Barbara E. "Requiem for the WASP," *Flying* 35 (Dec. 1944): 55–56.

Potter, Frank N. "The Swan Song Flight of Miss Mathilde," *Aviation Quarterly* 5 (Second Quarter, 1985): 182–188.

Powell, Hugh. "Harriet Quimby," *American Aviation Historical Society Journal* 24 (Winter 1982): 273–279.

Powers, Helen L. "From Hard Landings to Smooth Flights," *Women in Aviation* 2 (March-April 1991): 6–7.

Putnam, George Palmer. "Flyer's Husband," *Forum and Century* 93 (June 1935): 330–332.

Putnam, George Palmer. "Forgotten Husband," *Pictorial Review* 34 (Dec 1932): 12–13.

References

Putnam, George Palmer. *Soaring Wings: A Biography of Amelia Earhart* (New York: Harcourt-Brace, 1939).

"Queen Helen of the Air Breaks Down Industry's Last Barrier." *Literary Digest* (Oct. 26, 1935): 34.

Quimby, Harriet. "American Birdwomen — Aviation as a Feminine Sport," *Good Housekeeping* 55 (Sept. 1912): 315–316.

"Resumes Air Tours Urging Towns to Paint Names on Barn and Factory Roofs." *New York Times* (Aug. 31, 1947): II-10.

Rich, Doris L. *Amelia Earhart: A Biography* (Washington, D.C.: Smithsonian Institution Press, 1989).

Root, Amos I. "Amos I. Root Sees Wilbur Wright Fly," *Gleanings in Bee Culture* (Jan. 1, 1905): 36–39. Reprinted in Richard P. Hallion, *The Wright Brothers* (Washington, D.C.: National Air and Space Museum, 1978), 110–115.

Rothman, Sheila. *Women's Proper Place* (New York: Basic Books, 1978).

Rupp, Leila J. *Mobilizing Women for War* (Princeton, NJ: Princeton University Press, 1978).

"Ruth Alexander Death Discloses Doubts of Marital Status." *New York Times* (Sept. 20, 1930): 1.

"Ruth Elder." *Aero Digest* 13, no. 5 (August 1928): 238.

"Ruth Law and Her Remarkable Flight From Chicago to New York." *Scientific American* 115 (Dec. 2, 1916): 495.

"Ruth Law's Record Breaking Flight." *Flying* 5, no. 11 (Dec. 1916): 454–455.

"Says Her Speed Mark Was Not Broken." *New York Times* (May 18, 1951): 4.

"Says She Flew Under East River Bridges." *New York Times* (Oct. 22, 1928): 3.

Scharr, Adela Riek. *Sisters in the Sky, Vol. I* (St. Louis: Patrice Press, 1986).

Scharr, Adela Riek. *Sisters in the Sky, Vol. 2* (St. Louis: Patrice Press, 1988).

Seagrave, Sterling. *Soldiers of Fortune* (Alexandria, VA: Time-Life Books, 1981).

"Search Abandoned." *Time* 30 (July 26, 1937): 36.

Shea, Nancy. *The Air Force Wife* (New York: Harper, 1951).

Shea, Nancy. *The WAACs* (New York: Harper, 1943).

Shore, Nancy. *Amelia Earhart* (New York: Chelsea House, 1987).

Shuler, Marjorie. *Passenger to Adventure* (New York: Appleton-Century, 1939).

Shuler, Marjorie. "Their Home Is in the Troposphere," *Christian Science Monitor Magazine* (June 17, 1936): 5.

Sicherman, Barbara, and Carol Hurd Green. *Notable American Women: The Modern Period* (Cambridge, MA: Belknap Press, 1980).

Simmons, Margaret Irwin. *Sally Wins Her Wings* (New York: Thomas Y. Crowell, 1943).

Smith, Constance Babington. *Amy Johnson* (London: Collins, 1967).

Smith, Elinor. *Aviatrix* (New York: Harcourt Brace Jovanovich, 1981).

References

Smith, Elizabeth Simpson. *Breakthrough: Women in Aviation* (New York: Walker and Co., 1981).

Smith, Vi. *From Jennies to Jets* (Fullerton, CA: Sultana Press, 1974).

Southern, Neta Snook. *I Taught Amelia to Fly* (New York: Vantage Press, 1974).

Southgate, Barbara. "Not in Competition With Men," *Airwoman* 2 (Dec. 1934): 4.

"Stewardess on an Air-Liner." *Literary Digest* 115 (Feb. 18, 1933): 35–36.

Stewart, Oliver. *Aviation: The Creative Ideas* (New York: Praeger Publishers, 1966).

Stinson, Marjorie. "The Diary of a Country Schoolgirl at Flying School," *Aero Digest* 12 (Feb. 1928): 168–169.

Stockbridge, F. P. "She Took a Tip and Flew: Beulah Unruh," *American Magazine* 112 (Nov. 1931): 78.

Strother, Dora Dougherty. "The WASP Training Program," *American Aviation Historical Society Journal* 19 (Winter 1974): 298–306.

Studer, Clara. "Bread and Butter and Airwomen," *National Aeronautics* 13 (Oct. 1935): 20.

Taylor, John W.R. *Flight: A Pictorial History* (New York: Peebles Press, 1974).

Tate, Grover Ted. *The Lady Who Tamed Pegasus* (Bend, OR: Maverick Publications, 1984).

Thaden, Louise. *High, Wide, and Frightened* (New York: Stackpole and Sons, 1938).

Tiburzi, Bonnie. *Takeoff* (New York: Crown Publishers, 1984).

Uglow, Jennifer S. *The Continuum Dictionary of Women's Biography*, New Expanded Edition (New York: Continuum Publishers, 1989).

Underwood, John. *The Stinsons* (Glendale, CA: Heritage Press, 1976).

Valerioti, R. M. "Ruth Nichols," *American Aviation Historical Society Journal* 29 (Spring 1984): 27–40.

Veca, Donna, and Skip Mazzio. *Just Plane Crazy* (Santa Clara, CA: Osborne Publisher, 1987).

Vecsey, George, and George C. Dade. *Getting Off the Ground* (New York: E. P. Dutton, 1979).

Verges, Marianne. *On Silver Wings* (New York: Ballantine Books, 1991).

"Veteran Woman Stunt Artist Ends Own Life." *Los Angeles Times* (Dec. 21, 1945): 29.

Ware, Susan. *Holding Their Own: American Women in the 1930s* (Boston: Twayne Publishers, 1982).

"WASPs Volunteer Services to Army." Associated Press Dispatch (Dec. 20, 1944).

Wayne, Dorothy. *Dorothy Dixon Wins Her Wings* (Chicago, Goldsmith Publishers, 1933).

Weatherford, Doris. *American Women in World War II* (New York: Facts on File Books, 1990).

Wells, Helen. *Silver Wings for Vicki* (New York: Grossett and Dunlap, 1947).

References

Westin, Jeane. *Making Do: How Women Survived the 30s* (Chicago: Follett Publishing Co., 1976).

Wheeler, Ruthe S. *Jane, Stewardess of the Airlines* (Chicago: Goldsmith Publishers, 1934).

White, Ruth Baker. "The Sky's the Limit," *Independent Woman* 28 (Nov. 1949): 326–328.

Whyte, Edna Gardner, with Ann L. Cooper. *Rising Above It* (New York: Orion Books, 1991).

Willson, Dixie. *Hostess of the Skyways* (New York: Dodd-Mead, 1941).

"Wing Talk: Lady Bird." *Colliers* (May 10, 1947).

Woloch, Nancy. *Women and the American Experience* (New York: Alfred A. Knopf, 1984).

"Women as Flyers Considered." *New York Times* (June 7, 1921): 16.

"Women Fliers Act as Fashion Models." *New York Times* (March 16, 1935): 20.

"Women in Aviation." *New York Times* (Apr. 17, 1931): 22.

"Women in the Air." *New York Times* (July 22, 1939): 14.

"Women in the Air." *New York Times* (Sept. 1, 1930):12.

"Women Pilots? Some Say Yes, Some No, Some Maybe." *Newsweek* 6 (Nov. 16, 1935): 22.

"Women Test Pilots." *New York Times* (Nov. 19, 1943): 18.

"Women: Unnecessary and Undesirable." *Time* 43 (May 29, 1944): 66.

"World Speed Mark for Women." *New York Times* (May 13, 1951): 10.

Yeager, Jeana, and Dick Rutan, with Phil Patton. *Voyager* (New York: Alfred A. Knopf, 1987).

Young, David, and Neal Callahan. *Fill the Heavens with Commerce* (Chicago: Chicago Review Press, 1981).

"Youngest Flier in America a San Antonio Girl." *Aerial Age* 3, no. 5 (Apr. 17, 1916): 149.

Zophy, Angela Howard (ed.). *Handbook of American Women's History* (New York: Garland Publishing Inc., 1990).

Index

A and P rating, 226; defined, 239
Accidents, 16, 23, 24, 33, 34, 42, 53, 78, 190;
  in histories, 157; media criticism of
  women in, 89, 90, 91, 92; in war, 64, 67,
  211. *See also* Safety
Adams, Clara, 98
Adams, Jean, 111
Adventurism, 75, 92, 108, 157, 199
Aerial Age, 84
Aerial Age Weekly, 84, 85(photo)
Aerial ambulance services, 34
Aerial photography, 7, 157
Aerial Vagabond (Owen), 74
Aero, 84
Aerobatics, 14, 20, 22, 23, 25, 72, 86, 132,
  146; defined, 239. *See also* Stunts
Aero Digest, 90
Aeronautical drafting, 121
Aero Sportswoman, (Thomas) (*Popular Avia-
  tion*), 89
African-Americans, 23
Aikin, Stella, 35(photo)
Aircraft: women's access to. *See* Racing, mili-
  tary aircraft; Airwomen, aircraft
Aircraft carriers, 150, 150n, 156
Aircraft demonstrations, 17, 21, 61, 62–63,
  127, 143, 147, 149. *See also* Test pilots
Aircraft development, 11, 12, 16, 17–18, 26,
  27, 32, 33, 58, 79, 90, 152, 153, 156, 158
Aircraft manufacturing: employment of
  women, 28–29, 32, 51, 127, 198
Aircraft sales: by women, 28–29, 40, 44, 51, 57
Air Ferrying Command (U.S. Army Air
  Corps), 57
Air Force Academy, 210
Air Force Medal, 64
Airline Builders (Allen), 158
Airline pilots, 150, 173, 180, 181; advance-
  ment as group, 173, 177, 178, 181 (*see also*
  Leadership; Role models); background and
  motivation, 173, 174, 175, 176, 177; influ-
  ence of airwomen, 173, 174, 175, 178–
  179; in 1930s, 37, 38, 39, 102–103, 137

Airlines, 6, 98, 201. *See also* Airline pilots;
  Passenger service
Airmail, 18, 21, 38, 121, 192; in histories, 158
Air marking, 36–37; defined, 239. *See also* Na-
  tional Air Marking Program
Airport operation, 46
Airshows, 10, 11, 22, 24, 25, 37; in histories,
  156; post–World War II, 131–132
Air Transport Auxiliary (ATA) (Great Britain), 53, 57, 178, 179(table); Americans
  in, 38–39, 54, 57, 58, 127, 129,
  130(photo), 209–210; ATA girls, 53–54,
  110, 127, 144, 159, 224, 239; defined, 239;
  writings about, 110, 159, 224
Air Transport Command Ferrying Division
  (U.S. Army Air Corps), 57, 59, 61, 64, 66,
  127
Airwoman, 197
Airwomen, 5, 6, 7, 8, 13; advancement as
  group, 39, 143–144, 148, 149, 168, 189,
  201, 211, 225 (*see also* Individualism;
  Leadership; Professionalism); aircraft for,
  79, 186–189, 195; defined, 5, 6, 134 (*see
  also* Female pilots); in early popular culture, 170; entry into aviation, 14–15 (*see
  also* Career development); historical period of, 5, 6, 86, 116–118, 164, 165–167,
  168, 169, 170, 171, 172, 202–205; in-
  group loyalty, 67; present-day familiarity
  with, 178–179, 180–181, 183–184, 191,
  217–218; present-day recognition, 226–
  227, 228, 229, 233, 235, 236, 237–238 (*see
  also* Youth); and traditional roles, 134,
  196, 209, 212 (*see also* Family relations;
  Fashion; Gender roles); withdrawal of, 21,
  38, 42, 44, 81, 119–120, 125, 127, 129,
  131, 147–148, 189–191, 215; writings
  about, 75, 82, 84, 86, 97, 100, 104, 109,
  112, 224, 226 (*see also* Aviation histories;
  Biographies; Literature, juvenile; Media;
  Women's history); writings of, 70–77, 94,
  102, 145. *See also* Aviation; Airline pilots;
  Employment; Military pilots; *individual
  airwomen*

Index

Allegheny Airlines, 127
Allen, Brooke E., 136
Allen, Oliver E., 158
All-Woman Transcontinental Air Race (AW-TAR), 132, 133, 144, 146, 209; defined, 239
Altitude. *See* High altitude flight; *under* Record setting
Amelia Earhart Trophy, 196
America in the Air War (Jablonski), 158
American Airlines, 173
American Association of University Women, 197
American Heritage History of Flight (Gordon), 154–155
American Legion, 141
American Magazine, The, 82
American Society for the Promotion of Aviation, 32
American Women, 161
Amphibian aircraft, 157; defined, 239
Anderson, Opal, 54
Anti-aircraft training, 61
Army Air Forces, 206
Army Aviation Branch, 21
Arnold, "Hap," 56, 57, 58, 59, 63, 125, 193, 206, 208, 209, 211
Ascani, Fred, 212
Association of Women Airline Pilots, 234
Astronauts, 136, 210
ATA girls. *See under* Air Transport Auxiliary
Atlantic, 118
Atlantic Ocean crossings, 46, 47, 78, 89, 96; aircraft ferrying, 57 (*see also under* Ferry pilots); in histories, 157; solo, 33, 39, 47, 79, 81, 102, 109, 133, 220
Auriol, Jacqueline, 141, 142, 143, 144
Autobiographies, 70–76, 122, 145, 173, 219, 220, 222
Autogiro, 37, 47; defined, 239
Avenger Field (Sweetwater, Texas), 60, 209, 235, 236
Aviation, 5, 9, 11–12, 15; as entertainment, 10–11, 12 (*see also* Aerobatics; Performance; Stunts); historical period of (*see under* Airwomen); male arenas of, 176, 199, 201; and media, 71–72, 75, 86–87, 89; practical uses of, 26 (*see also* Commercial aviation); women's role in image, 27, 28, 29, 51, 61–63, 68, 92, 191–192, 194, 195 (*see also* Safety). *See also* Aviation education; Aviation history
Aviation Country Clubs, 33
Aviation education, 39, 122, 127, 133, 199, 200; curriculum development, 230; present-day, 228, 233. *See also* Flightschools; Pilot training

"Aviation for You and Me" (Nichols), 75
Aviation histories, 150–160, 180; airwomen in, 152, 153, 154, 155, 156, 157, 158, 159, 160, 180, 223. *See also* Biographical dictionaries
Aviation periodicals. *See under* Periodicals
Aviation: The Creative Ideas (Stewart), 153
Aviatrix, 153
AWTAR. *See* All-Woman Transcontinental Air Race

B- (military aircraft designation code), 239
B-17, 63, 211
B-19, 62
B-26, 62, 193
Bailey, Lady Mary, 157
Barnes, Florence "Pancho," 90, 138, 188; in film, 224–225; writings about, 158, 221–222, 227
Barnstormers and Speed Kings (O'Neill), 156
Barnstorming, 12, 22–26, 53, 86, 198; defined, 240, 241; in histories, 155, 157. *See also* Aerobatics
Batten, Jean, 155
Baty, Peggy, 227, 228, 229
Beard, M., 62(photo)
Beaverbrook, Lord, 58
Beech, Walter, 40, 188
Beech Bonanza, 126
Beechcraft, 29, 43, 44, 51, 187
Beechcraft Staggerwing, 188
"Behind the Ballyhoo" (Nichols), 75
Bell Aircraft, 125
Bellanca, Giuseppe, 32, 187
Bell P-39 Aircobra, 61–62, 193
Bell P-59, 163
Bendix, Vincent, 51(photo)
Bendix Corporation, 43
Bendix Trophy Race, 43, 188, 196, 211; in histories, 156; 1935, 43, 50; 1936, 38, 43–44; 1937, 51; 1938, 51, 52; post–World War II, 139, 140–141
Bera, Fran, 133
Berkin, Carol Ruth, 167
Bernheim, Molly, 133, 145
Bertinelli, Valerie, 225
Betzler, M., 62(photo)
Bibliographies (history of flight), 151
Bilstein, Roger E., 152–153
Biographical dictionaries: and airwomen, 161–164
Biographies, 111, 145, 147, 218–219, 221–222
Biplanes, 14, 17, 19, 30, 40, 41(photo), 43, 54, 79, 96; defined, 240
Bleriot, Louis, 16
Bleriot Type XI monoplane, 83(photo)

Bombers, 6, 62, 239; ferrying, 56–57, 63, 211. *See also individual aircraft*
Bone, Hugh, 186
Born to Love (film), 105
Boyer, Lillian, 14
Bragg, Janet Waterford, 227
Bridges: flying under, 22, 30, 87
Brinley, Mary Ann Bucknum, 212, 222
British Ferry Command, 57
Broadwick, Tiny, 13, 163
Brocksom, Guerdon, 137
Bromwell, Laura, 91
Brown, Dorothy M., 168–169
Brown, Marjorie, 73
Bucharest (Romania), 50
Burnham, Margaret, 107
Burry Port (Wales), 47
Bush pilots, 91, 158, 220
Bush Pilots (Time-Life Books), 158

C- (military aircraft designation code), 240
C-54, 122, 127
CAA. *See* Civil Aeronautics Authority
Career development, 14–15, 27, 28, 29–30; and performance publicity, 30, 33, 35; portrayed for youth, 110, 111, 146, 227, 229. *See also* Aviation education; Airline pilots; *under* Family relations
Cargo aircraft, 240. *See also individual aircraft*
Cargo pilots, 39, 53
Carl, Ann, 64, 163
Carlstrom, Victor, 18
Carmien, Amy, 226–227
Ceiling Unlimited: The Story of American Aviation from Kitty Hawk to Supersonics (Morris and Smith), 153–154
Central Airlines, 38
Cessna aircraft, 124
Chafe, William Henry, 166–167
Chamberlain, Clarence, 33, 188
Charter operations, 25
Chicago, 18, 24, 83
Christopher Strong (film), 105
Chronicle of Higher Education, 224
Civil Aeronautics Administration (CAA), 38, 55, 103, 122, 240
Civil Air Patrol, 34, 177
Civilian Pilot Training Program (CPTP), 54–55, 103
Clark, Frank, 105
Clark, Julia, 22
Cleveland (Ohio), 35, 41, 42, 43
Closed-course racing, 41, 132, 133, 140, 141, 142; defined, 240
Cobb, Jerri, 136, 210
Cochran, Jacqueline, 39, 43, 49–51, 51(photo), 52, 96, 132, 162, 179(table),

188, 189, 236; awards, 141; and fashion, 50, 51, 212, 213; in histories, 152, 154, 155, 157, 163, 164, 167, 170, 172; and jet aircraft, 142; in military piloting programs, 56, 57–58, 59–61, 125, 206–208; personal qualities and leadership, 205, 208, 209, 211, 212, 213, 214; post–World War II, 138–139, 139(photo), 140–144, 146, 148, 213; writings of, 209, 222. *See also* Odlum, Floyd
Cochran All-Woman Trophy Race, 132
Cody, Mabel, 25, 25(photo), 26; in histories, 156
Cole, Jean Hascall, 220
Coleman, Bessie, 23–24, 227
Colliers, 146
Combat pilots, 18, 20, 21, 22, 53, 54, 65, 120, 150, 186; denial of women as, 18, 21, 22, 103, 120, 150, 186, 190, 210; present-day, 229; in Soviet Union, 54; women instructors for, 20, 53, 65, 200, 201, 207; in writings, 74, 159
Commercial aviation, 18, 19, 24, 26, 39, 67, 68, 86, 192; in histories, 153; image of women in developing (*see under* Aviation); and media, 86, 103; postwar development, 123–124, 125, 126, 191–192; professional women pilots for, 27, 31–32, 33, 37, 72, 103, 104, 126. *See also* Airlines; Flying businesses
Concorde, 135
Continuum Dictionary of Women's Biography, The (Uglow), 162–163
Cooper, Ann L., 136
Corn, Joseph, 10, 192
Corpus Christi Naval Air Facility, 232
Cosmopolitan, 75
Country Life, 82
CPTP. *See* Civilian Pilot Training Program
Crosley, Powell, 188
Cross Country Derby, 42
Cross-country flying, 18, 33, 35, 41, 133
Crossen, Marvel, 91, 158
Curtis, Lettice, 144
Curtiss, Glenn, 22
Curtiss Carrier Pigeon refueler, 187
Curtiss Condor, 34
Curtiss Flying Service, 37
Curtiss Model E biplane, 19(photo)
Curtiss Pusher, 18, 22, 186
Cyrus, Diana, 140

Dailey, Janet, 223
Dare, Ethel, the Flying Witch. *See* Hobbs, Margie
Dayton (Ohio), 9, 234
deLeeuw, Hendrik, 154

Index

Democratic National Committee, 35
Democratic party electoral causes, 56
Depression. *See* Great Depression
Detroit (Michigan), 38, 44
Dippy twist, 20–21
Dirigible, 156; defined, 240
Discrimination, 38, 103–104, 134, 137, 197;
  in obtaining aircraft, 186, 187, 188, 195.
  See also Prejudice; Sexism
Distance record setting. *See under* Record
  setting
Ditching, 89, 122; defined, 240
Divebombing, 52, 61; defined, 240
Douglas, Deborah G., 223
Dover (England), 15
"Do Women Hoodoo Transatlantic Flights?",
  97
Dunlap, Vera Mae, 157

Earhart, Amelia, 6, 29, 32, 38, 39, 43, 44–45,
  45(photo), 46–49, 76(photo), 90, 107,
  122, 174, 187, 188, 196, 228, 236; in histories, 152, 154, 155, 157, 159, 160, 162,
  163, 164, 166, 168, 169, 170; and media,
  46–49, 77–79, 80(table), 81, 92, 93, 97,
  101; piloting skills, 32, 49, 78–79, 188,
  198; round-the-world flight, 48–49, 79,
  81, 92; writings about, 221, 224, 227; writings of, 75–77, 103. *See also* Putnam,
  George Palmer
Edwards, Rigdon, 235
Edwards Air Force Base, 158
Elder, Ruth, 87, 88(photo), 90; in histories,
  155
Embry-Riddle Aeronautical University, 227,
  228
Employment: gender-based hostility in postwar, 65–66, 100, 102, 103–104, 118, 137,
  168, 171, 191, 193, 207; postwar and present-day, 103, 106, 118–119, 124–127,
  132–133, 193, 202, 225; women in war-
  time, 116–117, 118, 202
Endurance flying, 30, 195; defined, 241. *See
  also under* Record setting
England. *See* Great Britain
English Channel, 15, 16, 54, 83
Epic of Flight (Time-Life Books), 155
Equal Rights amendments, 166
Erickson, Barbara, 64, 173
Exhibitions, 10, 16, 18, 19(photo), 20, 22, 37,
  53, 86, 120, 121; in histories, 154; teams,
  24, 25(photo), 26. *See also* Performance
Experimental Aircraft Association: annual fly-
  in, 11, 231, 237
Explorers, The (Jackson), 157–158

F- (military aircraft designation code), 241

F-86 Sabre, 142
FAA. *See* Federal Aviation Administration
FAI. *See* Federation Aeronautique
  Internationale
Fairchild Aircraft and Engine, 32
Fairfax Field (Kansas City), 129
Family relations, 196, 202, 210, 215, 216; in
  career development, 42, 46, 52, 57, 93,
  95, 111, 126; in histories, 155; media inter-
  est in, 92, 93, 95, 96–97, 100, 104, 195.
  See also individual airwomen
Farmer, James H., 105
Farnsworth Experimental Station, 54
Fashion, 13, 14, 15, 49, 50, 51, 100, 101, 104,
  112, 195, 197, 212, 213; media interest in,
  92–93
FBO. *See* Fixed-base operators
Federal Aviation Administration (FAA), 240
Federation Aeronautique Internationale
  (FAI), 31, 135, 195, 241
Female pilots (post–World War II), 131–138,
  145, 184; defined, 134, 135, 136; writings
  of, 135, 136, 137, 145. *See also* Airwomen
Feminism, 44, 49, 77, 109, 162, 165–166, 169,
  172
Ferrying Division, 66
Ferry pilots, 39, 53, 54, 56, 57, 59, 66, 126,
  129, 133; fighter aircraft, 64; gender-based
  experience difference, 125; transatlantic,
  57, 63, 186. *See also* Women Airforce
  Service Pilots; Women's Auxiliary Ferry-
  ing Squadron
Fighter aircraft, 6, 51, 52, 64, 141, 142, 189
Film, 104–106, 112, 129, 224–225, 233
First Aviators, The (Pendergast), 156
Fixed-base operators (FBO), 25, 133; defined,
  241
Flight Angels (film), 106
Flight: A Pictorial History (Taylor), 155
Flight attendants, 177, 228. *See also*
  Stewardesses
Flight chart case, 18
Flight financing, 48, 50, 89
Flight for Freedom (film), 106
Flight in America (Bilstein), 152–153
Flight instruction. *See* Flightschools; Pilot
  training
Flightschools, 20, 21, 24, 39, 46, 72, 132
Flying, 84, 104, 146
Flying boats, 33, 98; defined, 241
Flying businesses, 18, 20, 23, 33. *See also*
  Barnstorming
"Flying circuses," 24
"Flying Fortress." *See* B–17
Flying Hostesses (film), 106
"Flying is Fun" (Earhart), 75
Fokker Tri-Motor, 47

France, 15, 20, 24, 57, 141, 142
French Aeronautical Medal, 141
From Flying Horse to Man in the Moon
   (deLeeuw), 154
Frye, Charlotte, 29
Fundraising flights, 21
Gee Bee racer, 50, 188
Gender equality, 3–4, 12--13, 38, 58, 67, 68,
   86, 100, 205, 216; airwomen's writings on,
   72, 73, 76–77, 137, 219; and employment,
   102, 103, 104, 106; in media, 84, 90, 91,
   92, 96, 100, 101–104, 112; in military service, 53–54, 55, 58–59, 66, 103, 125–126,
   172; in politics, 203, 204; portrayal for
   youth, 107–111; post–World War II, 202.
   See also Gender relations
Gender relations, 72–73, 197, 203–204, 216;
   airwomen's writings, 73, 74, 76–77; and
   media, 92. *See also* Employment, gender-
   based hostility in; Gender roles; Sexism
Gender roles, 4, 5, 6, 7, 13, 30; adaptive responses to, 169, 170; airwomen's roles as
   unique, 90, 97, 98, 101, 134, 159, 172,
   207, 210, 212, 216 (*see also* Separateness);
   airwomen's writings on, 66, 71, 72, 74, 75–
   76; in film, 105–106; and media, 82, 86,
   87, 92, 93, 103, 112; and protection of
   women, 63, 92, 166, 186, 187, 209, 211
   (*see also* Progressivism); traditional, 12,
   171, 196, 197, 198, 201, 202, 203, 215;
   women as helpmates, 94, 95, 96, 100, 157,
   177 (*see also* Stewardesses); in World War
   II and postwar, 65, 116–118, 119, 134,
   171, 202, 214. *See also* Discrimination;
   Women's history
General aviation, 127; defined, 241
Gentry, Viola, 31, 195
Gillies, Betty Huyler, 63, 125, 126, 198, 211,
   215
Gillis, Fay. *See* Wells, Fay Gillis
Girl Aviators (Burnham) (juvenile fiction se-
   ries), 107–109
Gleanings in Bee Culture (journal), 10
Gloster jets, 54
Golden Eagle aircraft, 186
Gone to Soldiers (Piercy), 224
Good Housekeeping, 72
Goodwill flights, 34
Gordon, Lou, 47
Gower, Pauline, 58
Granger, Byrd Howell, 220
Granville Brothers aircraft, 50, 188
Great Britain, 16, 38, 53, 57, 63, 96, 102, 116,
   142, 157. *See also* Air Transport Auxiliary
Great circle route, 95
Great Depression, 34, 36, 102; in histories,
   169–170

Greene, Elizabeth, 158
Griffith, Lori, 175
Ground-loop, 48; defined, 241
Grumman Corporation, 103
"Gypsies." *See* Barnstorming

Haizlip, Mae, 39, 171, 188, 192
Haldeman, George, 89
Hallion, Richard P., 158
"Hall of Fame of the Air" (comic strip), 107,
   108(illus.)
Hamel, Gustav, 16
Handbook of American Women's History (Zo-
   phy), 164–165, 181
Happy Bottom Riding Club, 158, 222
Harbor Grace (Newfoundland), 47
Hardelot (France), 15
Harlow, Jean, 105
Harmon Trophy, 141, 146
Harris, Barbara, 166
Harris, Grace, 133, 135–136, 145
Harris, Helen Hodge, 227
Hart, Marion, 133, 135, 136
Hartmann, Susan M., 171–172
Heath, Lady Mary, 53, 89–90; in histories, 157
Helicopters, 156, 174
Hell's Angels (film), 105
Hepburn, Katherine, 105
Heroines of the Sky (Adams and Kimball), 111
High-altitude flight, 187. *See also under* Re-
   cord setting
High, Wide, and Frightened (Thaden), 74, 186
Hinckley, Robert H., 55, 200
History. *See* Airwomen, historical period of
Hlavacek, Jane Page, 140
Hobbs, Margie, 14
Hobby, Oveta Culp, 59, 206, 207
Holden, Henry M., 175
Holding Their Own (Ware), 169
Home Front and Beyond, The (Hartmann),
   171–172
Hondo Army Airfield (Texas), 129
Hostess of the Skyways (Willson), 110
Houston Municipal Airport
Hudson bomber, 211
Hughes, Howard, 105, 224
Humanitarian flight, 122
Husband-wife teams, 94, 95, 96, 192

IFR (instrument flight rules), 36
I Fly as I Please (Hart), 136, 145
"I Let my Daughter Fly" (Beatty), 97
Independent Woman, 203
Individualism, 6, 135, 136, 138, 144, 145,
   149, 169, 184
In-flight refueling, 43
Ingalls, Laura, 34, 44; in histories, 158

Intelligence, 36, 37, 54
Inter-City Airlines, 57
International Society of Women Airline Pilots, 173, 175
International Women's Air and Space Museum (IWASM), 233–234
Irvin Airchute Company, 32
IWASM. *See* International Women's Air and Space Museum

Jablonski, Edward, 158
Jackson, Donald Dale, 158
Jacqueline Cochran Cosmetics, 51, 212
Jane, Stewardess of the Airlines (Wheeler), 110
Japan, 21, 52, 117
Jet aircraft, 54, 64, 140, 141, 142, 163
JN-4 "Jenny," 23, 24, 198
Johnson, Amy, 53, 96; in histories, 155
Johnson, Martin and Osa, 157
Joyce Hartung Trophy, 196
Junkers JU 88, 54

Keil, Sally Van Wegenen, 223
Kenney (air chief), 209
Kidd, Ray, 137
Kimball, Margaret, 111
Krantz, Judith, 223

Lach, Helen, 14
Ladies Courageous (film), 106
Ladybirds (Holden and Griffith), 175
Lady Drummond Hay Memorial Trophy, 141
Lafayette Escadrille, 20
Laird aircraft, 188
Lansing, Elisabeth Hubbard, 110
LaRene, Jean, 39
Laroche, Baroness de, 156
Law, Ruth, 18–19, 19(photo), 39, 72, 120, 186, 189, 215; and family, 120–121, 190; and media, 83, 84; writings about, 152, 155, 227
Lea, Luanne C., 93
Leadership, 6, 71, 76, 77, 129, 135, 138, 143, 147–148, 149, 168, 205, 214, 215, 216; post–World War II societal, 203–204, 230; present-day, 173, 178, 226, 227; of WASPs, 206–208. *See also* Role models
Lemon, Dot, 145
Leslie's Weekly, 15
Lewis, Cathleen S., 151–152
Lewis, Dorothy Swain, 236
Liberty (publication), 208
Licensing, 15, 20, 23, 50, 156
Life, 93, 124, 146
Light, Richard and Mary, 157
Linda Carlton (Lowell) (juvenile fiction series), 109

Lindbergh, Anne Morrow, 93–95, 157, 192
Lindbergh, Charles, 12, 41, 47, 79, 93, 157, 192
Listen! The Wind (Lindbergh), 94
Literary Digest, 84, 89, 101
Literature, 7, 145, 147; juvenile, 107–111, 146–147, 224. *See also* Airwomen, writings of
Litvak, Lily, 54
Living Age, 82
Lockheed aircraft, 188
Lockheed C-5A, 232
Lockheed Electra, 48
Lockheed Hudson, 58
Lockheed Vega, 33, 47, 79, 188, 190
London, Terry, 173
London (England), 50, 96
Londonderry (Northern Ireland), 47
Long-distance travel, 94
Loops, 10, 18, 20, 72, 86, 91
LORAN, 36; defined, 241
Love, Nancy Harkness, 36, 127, 128(photo), 129, 179(table); military piloting program, 56–57, 59, 61, 63, 64, 172, 211; writings about, 227
Love, Robert, 57, 59, 127
Love Takes Flight (film), 106
Lowell, Edith, 109
Lowenstein-Wertheim, Princess Anne, 157
Loy, Myrna, 105

McCloskey, Helen, 36
McDonald, Katherine, 105
Mach, Mary von, 227
McKey, Alys, 22, 186
MacRobertson race, 50, 96
Manila (Philippines), 18
Mantz, Paul, 140
"Man Who Tells the Flier, 'Go,' The" (Earhart), 75
Mapes twins (Moyer) (in juvenile fiction), 109
Maps. *See* Flight charts
Marked course racing, 33
Markham, Beryl, 53, 102, 220–221
Marsalis, Frances, 43, 171
Martin, Glenn, 13
Martin B-26, 62
May, Charles Paul, 218
Mazzio, Skip, 222
Measured-course races, 40
Media: in airwomen's era, 5, 9–10, 15, 41, 70, 77–78, 82, 86, 89, 112, 119, 185; airwomen as correspondents, 32, 37, 70, 74, 75; and gender-related hostility, 100, 103 (*see also under* Sexism); post–World War II, 144–145, 199; present-day, 231; stereo-

types, 4, 82, 93, 108, 111, 158. *See also* Film; Literature; Popular culture; Press; Putnam, George Palmer; *individual air-women;* Aviation and media; Safety
Merrill, Dick, 102
Meyers, Dickey, 111
MGM (film corporation), 105
Mid-South Airways of Memphis, 25
Military academies, 210
Military aircraft: designation codes, 239, 240, 241, 242 (*see also individual aircraft*). *See also under* Racing
Military aviation histories, 153
Military pilots, 6, 20, 38–39, 52–54; British, 53–54, 56; post–World War II, 124; present-day, 229, 232; Soviet, 54, 56, 159; women's noncombat roles as, 55–57, 58–59, 67, 103, 150. *See also* Air Transport Auxiliary; Civilian Pilot Training Program; Women Airforce Service Pilots; Women's Auxiliary Ferrying Squadron; Women's Flying Training Detachment
Military pilot training, 53, 54–55, 58, 60, 61, 65, 132, 200, 207, 211, 216. *See also* Combat pilots, women instructors for Military service, 116–117, 150, 171–172, 213; combat requirements, 55–56, 103
Miller, Bernetta, 17, 156
Model airplanes, 107, 196
Moisant, Mathilde, 17(photo), 189; in histories, 154, 156; writings about, 227
Moisant International Aviators, 16–17, 17(photo); in histories, 154, 156
Mollison, James, 96
Mono Aircraft Corporation, 188
Monocoupe racers, 188
Monoplanes, 15, 17, 19, 42, 76(photo), 79, 83(photo), 188
Moolman, Valerie, 159
Morale-boosting, 63
Morris, Lloyd, 153–154
Moyer, Bess, 109
Multi-engine aircraft, 19
Museums, 152, 165, 223, 233, 234

Nagle, Nadine Canfield, 227, 233
NASA. *See* National Aeronautics and Space Administration
Nation, The, 82 , 89
National Advisory Committee for Aeronautics: Special Adviser for Air Intelligence, 36
National Aeronautics, 27
National Aeronautics and Space Adminstration (NASA), 210
National Air and Space Museum, 152, 165, 223

National Air Marking Program, 36, 43, 57, 132, 134–135, 190, 216
National Air Races, 42, 43, 91
National Geographic Society, 95
National Women in Aviation Conference, 227–230
National Women's Air Derby: 1929, 41–42, 44, 79, 90–91, 132, 188; in histories, 152
National Women's Party, 49, 166
Navigation. *See* Air marking
Needed: Women in Aviation (Meyer), 111
Nevin, David, 157
Newark (New Jersey), 47, 98
New Deal, 6
Newsweek, 103
New Wings for Women (Knapp), 147
New York, 18, 33, 41, 43, 51
New York Times, 79, 81, 82–83, 84, 87, 91, 96, 101, 102, 103, 121, 127, 145, 153
Nichols, Ruth, 29, 30, 31, 32–34, 46, 47, 120, 122–123, 146, 188, 189, 190, 204; in histories, 152, 153, 155, 162, 171; writings of, 74–75, 103
Night flying, 18
99er (aviation publication), 197
Ninety-Nine News, 197
Ninety-Nines, 121, 131, 132, 134, 136, 138, 144, 149, 184, 209, 229; establishing niche, 199, 205; in histories, 164; publications of, 197. *See also* Museums
Noggle, Anne, 224
Noonan, Fred, 48
Northrop Gamma, 51, 188
North to the Orient (Lindbergh), 94
Northwest Airlines, 230, 231
Norton, Mary, 203
Norton, Mary Beth, 167
Notable American Women (James; Sicherman), 162
"Not in Competition With Men" (Southgate), 199
Nova Scotia, 102
Noyes, Blanche, 29, 36, 37, 43, 132, 134, 146, 205

Oakes, Claudia, 165
Oakland (California), 34, 47, 48
Odlum, Floyd, 50, 51, 52, 142; military piloting program, 56, 57
O'Donnell, Gladys, 79, 107, 192
Olds, Robert, 56, 57, 58, 59
O'Malley, Patricia, 110
Omlie, Phoebe Fairgrave, 24–25, 35, 35(photo), 36, 120, 121, 188, 191, 192, 200, 201, 216, 228; in histories, 162
Omlie, Vernon, 25, 192
O'Neill, Lois Decker, 163

Index

O'Neill, Paul, 156–157
O'Neill, William N., 165–166
Osborne, Carol, 222
Owen, Bessie, 74, 134

P- (military aircraft designation code), 241
P-35, 52
P-38, 67, 140
P-39, 61–62, 193
P-47, 52, 64, 211
P-51 Mustang, 139, 139(photo), 141, 211
P-59, 64
P-63, 125
Pancho Barnes (television film), 224–225
Parachute jumping, 10, 13, 14, 24–25, 163, 200
Parks College (St. Louis University), 227
Passenger service, 18, 19, 28, 33, 72, 192
Passenger travel, 98, 101, 121; in histories, 156, 157, 158
Pateman, Yvonne (Pat), 232
Pathfinders, The (Nevin), 157
Patriotism, 64, 67, 111, 117, 194, 200
Patterson, Robert, 209
Pay scales, 59
Pendo, Stephen, 106
Pennsylvania School of Aviation, 42
Pepper, Claude, 203
Percival Mew Gull, 102
Performance, 11, 12, 13–14, 18, 26, 27, 66, 67, 68; in histories, 156; image of, 86. *See also* Barnstorming; Exhibitions; Record setting; Showmanship
Periodicals, 82–84, 146; aviation, 84, 89, 146, 197, 199, 218, 225–227
Piercy, Marge, 224
Pilots. *See* Women pilots
Pilot training, 20, 54
Pilots' union, 103
Piper aircraft, 124
Pisano, Dominick A., 151–152
Pittsburgh Aviation Industries Corporation, 42
Political activity, 35, 203–204
Popular Aviation, 89
Popular culture, 160, 169, 170, 172, 180. *See also* Media
Post, Wiley, 37, 48, 197; in histories, 157
Postage stamps, 236–237
Powder-Puff Derby, 132
Prejudices, 44, 184; writings about, 71, 74, 77, 104, 147. *See also* Discrimination; Sexism
Press, 70, 78, 79, 80(table), 81, 82, 95; editorial comment, 101, 102, 146; post–World War II, 145–146. *See also New York Times*; Periodicals

Professionalism, 4–5, 28; conferences, 227–229, 230; and group advancement, 143–144, 148, 149, 230; organizations for, 204, 225, 232; periodical, 197; post–World War II, 117, 134, 148, 204; present-day, 150, 176, 225; in society, 166
Professional piloting, 4–5, 72, 111, 190, 192, 215; airwomen's writings about, 72; and performance, 27, 31, 39, 67. *See also* Airline pilots; Airwomen; Military pilots; Racing; Test pilots; *under* Commercial aviation
Progressivism, 6, 12–13, 86, 165, 166, 167
Prohibition, 6, 165
Protection of women. *See under* Gender roles
Public interest: airwomen's era, 5, 9, 10, 18, 23, 27, 28, 40, 41, 52, 67, 79, 87. *See also* Press
Publicity: airwomen's era, 33, 38, 39, 46, 47, 79
Public relations, 21, 43
Purdue University: Research Foundation, 48; Women's Careers Department, 48
Pursuit aircraft, 241
Putnam, George Palmer, 32, 46, 47, 48, 49, 78, 79, 81, 97, 101, 188
Pylon racing. *See* Closed-course racing

Quimby, Harriet, 15–16, 17(photo), 70, 83, 83(photo), 90, 109, 143, 179(table), 236; in histories, 152, 155, 156, 164; writings about, 227; writings of, 71

Racing, 27, 29, 35, 37, 38, 39, 47, 86, 90; exclusion of women, 43, 196; military aircraft, 139, 140, 142, 187; post–World War II women's, 131, 132, 133, 135, 138, 139, 195, 196; transcontinental, 5, 35, 41, 42, 43, 79; women's events, 131, 195, 196; writings about, 75, 109. *See also* Closed-course racing; Cross-country flying; Flight financing; Record setting; *individual races*
Radcliffe College, 162
Ramsey, Nadine, 140
Randolph Air Force Base, 232
Record setting, 27, 29, 31, 37, 38, 66, 83, 86, 105, 131, 186; altitude, 26, 31, 33, 40, 47, 52, 86; distance, 18, 21, 26, 33, 34, 86; endurance, 21, 26, 31, 40, 43, 86; as passengers, 98; speed (jet), 140, 142; speed (propeller), 21, 26, 31, 33, 40, 44, 51, 52, 86, 140, 141; women's classifications, 31, 186, 195, 196; writings about, 75, 159. *See also* Racing
Relief Wings (aerial ambulance service), 34
Republic P-35, 52
Republic P-47 Thunderbolt, 52, 64

Richards, H., 126(photo)
Richey, Helen, 36, 37–39, 58, 101, 102, 129, 130(photo), 179(table); in histories, 170
Rickenbacker, Eddie, 107
Ride, Sally, 163
Rising Above It (Whyte and Cooper), 136
Risk-taking, 87, 90, 91
RKO (film corporation), 105
Roberts, Beatrice, 106
Rogers, Will, 37, 197
Rogers Airlines, 33
Role models, 31, 39, 66, 72, 77, 138, 173; in histories, 162, 163, 168, 170, 172; present-day, 177–178, 181, 226, 228
Romulus (Michigan), 62, 125
Roosevelt, Eleanor, 55, 56, 58
Roosevelt, Franklin Delano, 35, 36, 56
Roth, Vita, 200
Rothman, Sheila, 167
Round-the-world flight, 37, 46, 48, 78, 81, 98, 105, 174, 188; in histories, 157
Roy, Gladys, 14, 156
Royal Canadian Flying Corps, 20
Rutan, Dick, 174
Ruth Chatterton Trophy, 196
Ryan Aeronautical, 126

Safety, 17, 24, 27, 61–62, 84, 199; and media, 86, 90, 91, 92
Sally Wins Her Wings (Simmons), 109
San Antonio (Texas), 20
Santa Monica (California), 35, 41, 42
Saturday Evening Post, 124, 146
Saturday Review, 145
Schamberger, Kate, 233
Scharr, Adela Riek, 221
Scientific American, 83
Scott, Blanche, 22, 146, 227, 236
Seagrave, Sterling, 159
Seaplanes, 33, 47, 157
Searchlight tracking, 61
Seatbelts, 17, 24, 91
Second Ferrying Group, 125
Separateness, 92, 194–197, 199–201, 204, 215
Seversky, Alexander de, 51, 51(photo), 52, 189
Seversky Pursuit, 51(photo)
Sexism, 3–4, 14, 38, 51, 60, 68, 87, 92, 150n, 153, 203; in career development, 28–30, 100, 111, 137, 184, 189, 215, 216; and media, 82, 97, 99–101, 103, 106, 159; in war effort, 57, 62, 63, 64, 65–66, 193. *See also* Discrimination; Prejudice
Sharp, Evelyn, 67
Showmanship, 10, 11, 23, 66; in histories, 156
Silver Wings for Vicki (Wells), 146
Silver Wings Santiago Blue (Dailey), 223

Simmons, Margaret Irwin, 109
Skelton, Betty, 146
Sky Service (Lansing), 110
Smith, Elinor, 29, 30–32, 47, 87, 120, 122, 187, 188, 196, 215; writings of, 219
Smith, Kendall, 153–154
Smith, Tom, 30
Snook, Neta. *See* Southern, Neta Snook
Societal trends. *See* Airwomen, historical period of
Solo flights, 40, 41, 47, 81, 90, 96, 102, 106; in histories, 157
Southern, Neta Snook, 45(photo), 46, 198, 215
Southgate, Barbara, 199
Soviet Union, 37, 53, 209
Space program, 233. *See also* Astronauts
Spartan Aircraft, 39
Spartan School of Aeronautics, 132
Speed records. *See under* Record setting
"Sportsman Flies His Plane, The" (Nichols), 74
Stars at Noon (Cochran), 145, 171
Steadman, Beatrice, 233
Stereotypes. *See under* Media
Stewardesses, 98–100, 101, 106, 184, 195; in histories, 153, 155, 158, 160, 180; in juvenile fiction, 110–111, 146. *See also* Flight attendants
Stewart, Oliver, 153
Stinson (aircraft manufacturers), 19, 124; in histories, 155, 164
Stinson, Edward, 20
Stinson, Katherine, 20–21, 22, 31, 85(photo), 120, 121, 179(table), 189, 190; in media, 72, 84; writings about, 152, 154, 227
Stinson, Marjorie, 20, 21, 72, 121; in histories, 154; in media, 84
Strafing, 61
Stranger Than Fiction (film), 105
Stratemeyer (general in armed forces), 209
Strohfus, Elizabeth, 230–231
Stultz, Wilmer, 47
Stunts, 5, 14, 23, 30, 224; media reporting, 87
Suffrage, 165, 166, 203
Suicides, 123, 129, 131
Supercharging, 32; defined, 242
Sweetwater (Texas), 60, 235, 236

T- (designation code), 242
T-6, 133, 144, 145
Taboos, 97
Target-towing, 61, 230
Tate, Grover Ted, 221–222
Taylor, John, 155
Television. *See* Film
Tennessee: flight training program, 200, 201

Tennessee Aviation Commission, 122
Test Pilot (film), 105
Test pilots, 17, 22, 32, 57, 61, 64, 105, 199; in histories, 158; and media, 103–104; post–World War II, 126, 141, 144
Texas Escadrille, 20
Thaden, Herbert Von, 42
Thaden, Louise McPhetridge, 29, 31, 36, 39, 40, 41, 41(photo), 42–44, 51, 79, 120, 122, 162, 179(table), 187, 189, 190, 211, 215; and family, 42, 43, 44; in histories, 157, 170, 171; writings about, 227; writings of, 73–74, 186
Thaden T-4, 42–43
Thomas, Joan, 89
Tiburzi, Bonnie, 173–174
Tier, Nancy Hopkins, 227, 233
Till We Meet Again (Krantz), 223–224
Time, 92
Time-Life Books, 155–160
Too Hot to Handle (film), 105
Towing, 61, 230
Training aircraft, 126(photo), 242. *See also* T-6
Transatlantic crossings. *See* Atlantic Ocean crossings
Transcontinental air races. *See under* Racing
Transcontinental Handicap Air Derby, 35
Transportation industry, 11; development, 86, 90, 102
Transportation routes, 5, 98
Transport pilots, 6, 38, 199, 200. *See also* Airline pilots
Trapeze acts, 10, 14
Travel Air Company, 29, 40; racing planes, 40, 41(photo), 42, 187. *See also* Beechcraft
Travel Air Mystery Ship, 188
Trepassy Bay (Newfoundland), 47
Trout, Bobbi, 31, 186, 191, 222, 228
Truman, Harry S., 141
Tunner, William, 59, 63, 64, 193
Two Hundred Years of Flight in America: A Bicentennial Survey (Emme, ed.), 153

Uglow, Jennifer S., 162
UNICEF, 122
United Air Transport, 98
United States, 53
United States Women in Aviation (Douglas), 223
Unruh, Beulah, 97
U.S. Air Force, 212–213
U.S. Airmail Service, 21, 121, 190
U.S. Army, 17, 51
U.S. Army Air Corps, 57
U.S. Astronaut Program, 210
U.S. Postal Service, 236–237

Vampire fighter plane, 141
Vandenberg, Hoyt, 213
Veca, Donna, 222
Verges, Marianne, 223
VFR (visual flight rules), 36
Volunteer pilots, 20, 34, 125, 199
VOR, 36; defined, 242
Voyager aircraft, 174

WAAC. *See* Women's Army Auxiliary Corps
Waco biplanes, 30, 158
WACO Model 9, 88(photo)
WAC. *See* Women's Army Corps
WAF. *See* Women in the Air Force
WAFS. *See* Women's Auxiliary Ferrying Squadron
Warbird, 140; defined, 242
Ware, Susan, 169
Washington, D.C., 21, 38
WASPs. *See* Women Airforce Service Pilots
Wedell-Williams aircraft, 188
Wells, Fay Gillis, 29, 37
Wells, Helen, 146
Westin, Jeane, 170
West to the Sunrise (Harris), 135, 136, 145
West With the Night (Markham), 102, 178, 179(table), 220–221
WFA. *See* Women Flyers of America
WFTD. *See* Women's Flying Training Detachment
Wheeler, Ruthe S., 110
Who's Who in America, 161
Who's Who of American Women, 161
Whyte, Edna Gardner, 39, 132–133, 136–138, 145; writings about, 227; writings of, 219
Whyte, Murphy, 138
Wiggins, Mary, 129, 131
Willson, Dixie, 110
Wilmington (Delaware), 60, 125
Wilson, Woodrow, 18
Wing loading: defined, 242
Wing walking, 10, 14, 82; defined, 242; in histories, 157, 160
Woloch, Nancy, 168
Women Airforce Service Pilots (WASPs), 6, 61, 62(photo), 63, 64, 65–66, 67, 68, 102, 106, 126, 193(photo), 128(photo), 178, 179(table), 186, 209, 211; aircraft demonstrations, 61–63; denial of transatlantic ferrying, 63, 186; Director of Women Pilots, 61, 206, 213; disbanding and withdrawal, 65–66, 103–104, 119, 124–126, 127, 129, 131, 189, 191, 193, 208; and family, 125, 191; and fashion, 93, 212; in histories, 152, 153, 154, 158, 159, 163, 164, 172, 219–220, 223, 224; and media, 103–104, 223–224, 225; military benefits,

126, 223; nonmilitary status, 65, 125, 206, 207–208; opposition to, 66, 68, 193–194, 207; organization of, 60–61, 206, 207, 208; present-day publicity and recognition, 228, 230, 231, 232, 235–236; volunteering, 125, 199; wishing well of, 235, 236. *See also individual women*
Women Aloft (Moolman), 159–160, 181
"Women and Courage" (Earhart), 76
Women and the American Experience (Woloch), 168
Women Flyers of America (WFA), 199–200
Women in Aviation (Peckham), 146
Women in Aviation (periodical), 225
Women in the Air Force (WAF), 213
Women Military Aviators, 232
Women of America: A History (Berkin and Norton), 167
Women of Courage (video documentary), 225
Women pilots. *See* Airline pilots; Airwomen; Female pilots (post–World War II); Ferry pilots; Military pilots; Test pilots
"Women's Activities" (Heath) (*Popular Aviation*), 89
Women's Air Derby. *See* National Women's Air Derby
Women's Air Reserve (flying club), 129
Women's Army Auxiliary Corps (WAAC), 59
Women's Army Corps (WAC), 172, 206, 208
Women's Auxiliary Ferrying Squadron (WAFS), 59, 60, 61, 63, 127, 242
Women's Book of World Records and Achievements (O'Neill), 163–164
Women's Flying Training Detachment (WFTD), 60–61

Women's history: and airwomen, 165, 166, 167, 168, 169, 170, 171, 172, 180; studies of, 151, 160, 165–172, 180, 181. *See also* Airwomen, historical period of
Women's Home Companion, 118
Women's International Aeronautic Association, 141
Women's liberation, 73
Women's National Air Meet: 1934, 37, 196
Women's Research Flight Instructor School, 201
Women's roles. *See* Gender roles
Woods, Jessie, 227
Workplace reform, 118, 169
Works Progress Administration, 36
World Flight (humanitarian organization), 122
World War I, 18, 20, 52
World War II, 6, 34, 36, 38, 52–54, 58, 65; in histories, 158, 159, 166–167, 171
Wright, Orville and Wilbur, 9–10, 234; in histories, 156

Yeager, Chuck, 142, 158, 229
Yeager, Jeanna, 163, 174, 176
"You Must Fly" (Nichols), 75
Young, Cheryl, 230, 231
"Your Next Garage May House an Autogiro" (Earhart), 75
Youth, 107, 112, 227; literature for, 107–111. *See also* Aviation education

Zophy, Angela Howard, 164
Zuchowski, J., 62(photo)